Biotech's

# DICTIONARY *Of* Agronomy

*Compiled & Edited by*

JB Jain & Sumit Jain

BIOTECH

2005

BIOTECH BOOKS

DELHI - 110035

Biotech's

# DICTIONARY *Of* Agronomy

*Compiled & Edited by*
**JB Jain & Sumit Jain**



First Indian Edition 2005


ISBN 81-7622-142-2

*Published by* : BIOTECH BOOKS
1123/74, Tri Nagar,
Delhi - 110 035
Phone: 27382765
e-mail: biotechbooks@yahoo.co.in

*Showroom* : 4762-63/23, Ansari Road, Darya Ganj,
New Delhi - 110 002
Phone: 23245578, 23244987

*Printed at* : Chawla Offset Printers
Delhi - 110 052

PRINTED IN INDIA

# PREFACE

Agronomy is a branch of agriculture dealing with various physical and biological factors—including soil management, tillage, crop rotation, breeding, weed control, and climate—related to crop production. Agronomy commonly refers to field crops, e.g. wheat, rice, corn, sorghum, soybean, cotton, as well as pasture, sugar, and forage crops, while horticulture is concerned with fruits, vegetables, flowers, and ornamental plants, silviculture or forestry – with forest trees, agroforestry – with mixtures of trees with other crops.

Most of the words and abbreviations are not to be found in a standard collegiate dictionary and many are not present in the dictionaries of biology. This is because many of the terms are newly coined. Such dictionaries like this have fascinated the teachers and taught everywhere. They stimulate us in knowing the unknown. In fact, a dictionary of such a kind has an ocean of information which tries to satisfy our unending thirst for bio-physical knowledge.

The task of selecting the terms in any branch of science offers many difficulties, particularly in the case of agriculture. How far it is advisable to include terms from those overlapping sciences, which lie on the border land is the question, on which no two people might think alike. Every word has been given an independent examination, so as to take in all, in fact, which might be fairly excepted, and yet to exclude technical terms which really belong to another science, which otherwise would have extended this dictionary to an inconvenient size.

▪ **a bullock work unit**
the average work put up by a single bullock in eight hours.

▪ **$^{12}$c$^{13}$c ratio**
a ratio of the proportion of $^{12}C$ to $^{13}C$ in a geological deposit which determines (whether or not) it is of living matter.

▪ **A**
A symbol of mass number. It is the sum of Z (atomic number) and N (neutron number).

▪ **a c soil**
in this type of soil the transition is formed as a humic horizon. This horizon is not detected through a microscopic examination. It is also less highly developed than an AC soil.

▪ **a frame ridger**
an implement of wood made for making bunds and ridges. This helps in extending the existing bunds and gathering soil from a wider strip.

▪ **a(b)c soil**
a type of soil with three horizons in which, the upper A horizon is differentiated from B primarily by weathering.

▪ **a2 horizon**
a mineral horizon in which the feature emphasised is loss of clay, iron or aluminium, with resultant concentration of quartz or other resistant minerals in sand and silt sizes. The main feature is lighter colour and lower organic content.

▪ **a3 horizon**
a transitional layer of minerals between A and B, and dominated by properties characteristic of an overlying AI and Az but having some subordinate properties of an underlying B.

▪ **ab horizon**
an intermediary horizon between horizon A and B. The upper part of this horizon constitutes the properties of horizon A and the lower part constitutes the qualities of the horizon B, and the two parts cannot conveniently be separated into $A_3$ and bi.

▪ **abatement**
it is the method that is used for reducing the intensity of pollution, be it air, water or noise.

▪ **abc soil**
a type of soil where there are three layers A, B and C of soil visible.

▪ **abd soil**
a type of soil where the parent rock has been altered entirely.

▪ **aberrant**
a plant or soil that differs from the normal characteristics of a specie, genus, or higher group in one or more characters.

▪ **aberration**
when a plant develops a peculiarity usually due to some special environmental factor but the characteristics are not visible in the offsprings.

▪ **abiotic environment**
that part which consists of the topography, climatic factors, and inorganic nutrients of the environment.

▪ **abnormal**
refers to either above or below the normal value of all element, regardless of the degree of deviation.

▪ **abnormal seedlings**
the abnormal characteristics that a seedling of a healthy normal plants shows of discontinued growth while grown in good quality soils under favourable conditions of water supply, light and temperature.
Examples are seedlings with weak or malformed parts such as stubby roots, split hypocotyl, absence of primary leaves and/or terminal bud.

▪ **abnormal soil**
the soil that shows abnormal characteristics that develop within the normal decree of the climate considered or when other factors have intervened.

▪ **aboriculture**
when an individual tree is cultivated thoroughly for fruits, mats, gums, resins etc.

▪ **abrasive**
is the condition when abrasion is produced.

▪ **abrupt textural change**
the boundary that differentiates the ochric epipedon or albic horizon and an underlying argillic horizon. When there is ochric <20% clay in epipedon or albic horizon, then there will be double the contents in 7.5 cm or less. But if there is a >20% of clay in the ochric epipedon or albic horizon, an increase of 20% clay of absolute clay content is required in 7.5 cm or less.

▪ **abscisic acid (aba)**
it is an important constituent in mediating the adaption of a plant to stress. The physiological effects-inducing promotion such as senescence and abscission, causing dormancy in buds and seeds, closure of stomata, retardation and inhi-

bition of growth are the results of this acid. It is also known as 'antigibberellin' since it inhibits gibberellin stimulate of growth.

- **abscission**
when the parent plant sheds the leaves, fruits or stems, the process is called abscission.

- **absolute**
not relative or independent

- **absolute atmosphere**
when the absolute pressure equal to one million times the pressure is produced on a square centimetre by the force of one dyne, it is called absolute atmosphere.

- **absolute deviation**
the difference between the two measurements or an observation and a measurement such as the mean used as a point of origin or reference.

- **absolute drought**
when for at least consecutive fifteen days there has been no or not more than 0.25 millimetres of rainfall.

- **absolute growth rate**
under a given time and specific condition, the rate of growth of a plant is called the absolute growth rate.

- **absolute humidity**
in the unit volume of water, the mass of water vapour present is called the absolute humidity. Comparatively, cold air has a low absolute humidity than warm air. Its unit is expressed in gms/m$^3$ or in gms/ft$^3$.

- **absolute temperature**
the temperature that is used in the meteorological studies based on absolute zero (- 273°C), which is also known as O° Absolute or Kelvin (K). it is the point where the thermal motion ceases.

- **absolute transpiration**
when the rate at which the plant looses water is determined, the process is called the absolute transpiration.

- **absolute water requirement**
the amount of water absorbed by a crop in acre-inches per crop-season together with the evaporation from the crop producing land. It is also known as the consumptive use of water.

▪ **absolute zero**
it is the temperature at which all the thermal motion of atoms and molecules cease. The unit of absolute zero is K(Kelvin) = -273.18°C.

▪ **absorbance**
the process of transmission or reception of a particular quantity of a pure solvent.

▪ **absorbed water**
the amount of water absorbed by soil that is mechanically held in a soil mass. Its properties are not substantially different from ordinary water.

▪ **absorbency**
1. to take in and incorporate.
2. assimilate

▪ **absorbent**
when a liquid or a gas is taken up and distributed by a substance throughout its mass. For e.g., charcoal and purified cotton.

▪ **absorber**
when a material is specifically used for the absorption of the radioactive rays, it is called an absorber. Absorbers are used:
1. for the determination of the characteristics of a specific radiation.
2. as shields for reducing the intensity of radiation for safe handling and storage.
3. for removing a particular component from a complex source of radiation. Calibrated absorbers are usually made in the form of metal discs.

▪ **absorption**
the entrance of water into the soil or rocks by all natural processes. It includes the infiltration of precipitation or snowmelt, gravity flow of streams into the valley alluvium into sinkholes or other large openings, and the movement of atmospheric moisture.

▪ **absorption loss**
at the time when water is first turned into the reservoirs and canals, the water that is lost from the structure by absorption by the soil.

▪ **absorption radiation**
a process in which all or part of the energy of incident radiation is transferred to the mat-

place over a specified period. This determines whether an area is suitable for the cultivation of particular crops.

- **accumulation or summation curve**
  a curve showing the accumulative percentage by weight of particles within increasing or decreasing size limits as a function of diameter, the percentage by weight of each size fraction is plotted accumulatively on the ordinate as a function of the total range of diameters represented in the sample plotted on the abscissa.
- **acentric**
  not arising from the centre.
- **acethion**
  is an analogue of malathion having low mammalian toxicity. Its chemical formula is O, O-diethyl-S-carbothioxymethyldithiophosphate.
- **acetylase**
  an enzyme that catalyses the formation of acetyl esters.
- **achlorophyllous**
  a plant that lacks chlorophyll.
- **acid anhydride**
  an acid with one or more mol-





ter through which it passes by various interactions with the basic particles (electrons, nuclei, etc.) of which the matter consists. This process results in a reduction of the radiation intensity or a reduction in the number of incident particles. Also, a nuclear reaction or process in which a particle is absorbed by a nucleus, frequently resulting in the release of a different particle by the nucleus.

- **absorption ratio**
  the ratio between the final internal concentrations to the final concentration of a particular ion in plant tissue. It is denoted as *i/e*.
- **abundance**
  when there is an access of individuals of a given species as compared to the total number of individuals of the group considered.
- **ac horizon**
  it is the transitional horizon between A and C horizons of soil. This layer has the having underlying properties of both A and C layers, but it does not show the distinct qualities of any of the two layers. This nomenclature is used when the A-horizon rests on C-horizon.
- **ac soil**
  it is a type of soil where the two horizons, A and C, are clearly visible. In this type of soil the A horizon is formed from physical weathering, loosening and build up of organic matter of C horizon and not by migration from other type of soil.
- **accelerated erosion**
  soil erosion is the result of the activities of man. For example, preparation of land for the production of food and fibre and cutting down of trees for the purpose of building houses and for the purpose of building society etc. The action of wind and water are the primary causes of accelerated erosion.

- **accelerator**
  any substance used to assist the decomposition of organic matter in the preparation of garden compost.
- **acceptability, animal**
  the qualities that an animal

uses to judge a food item a
forage.

- **accessory**
auxiliary.

- **accessory pigments (photosynthesis)**
pigments like carotenoids a
phycobiliproteins that p
duce light energy
photosythethic process i
plant.

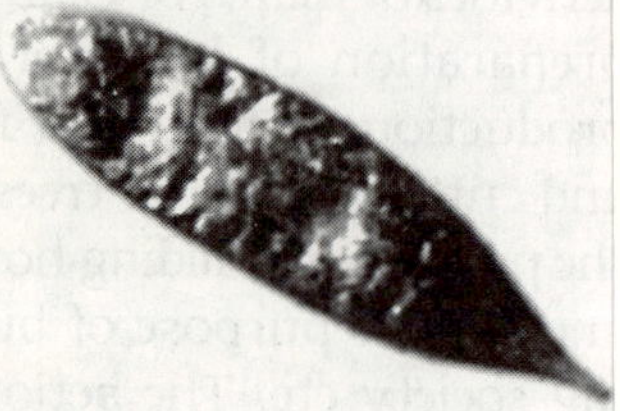

- **accidental error**
an error for which it is equa
probable that the sign of t
error is plus or minus. In
series of measurements, ac
dental errors are compensat
in their total effect.

- **acclamatory response**
when an organism brin
about reversible change in t
morphology or physiology
response to environment
change.

- **acclimation**
the physiological a
behavioural adjustments th
an organism brings about
produce changes in its imm



metrically how much of a standard solution is required to produce a given reaction.
2. refers to a titration in which an acid of known concentration is added to a solution of base of unknown concentration, or the converse.

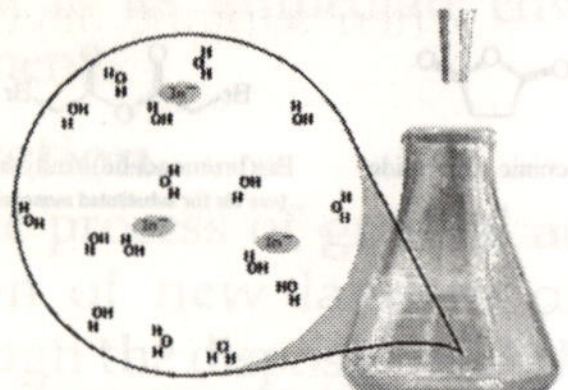

$HIn + OH^- \rightarrow H_2O + In^-$

- **acid burgundy mixture**
this mixture is a result of reaction between less than 0.86 parts of crystalline sodium carbonate with one part of copper sulphate.

- **acid electrolyte**
the acidic solution formed when a compound, such as sulphuric acid, conducts an electric current that dissociates into ions when dissolved.

- **acid equivalent**
refers to the portion of the formulation that theoretically can be converted to the acid.

- **acid fast bacteria**
when a bacteria after being stained with a hot dye such as carbol fuchsin strongly resist decolouration with acid alcohol. For example, *Mycobacterlum tuberculosis*.

- **acid fermentation**
when sewage sludge decomposes anaerobically, which results in the biodegradation of complex organic compounds to simpler organic acids, as in anaerobic sludge digestion.

- **acid former**
1. a group of facultative and obligate anaerobes, capable of hydrolysis, which ferment the complex organic compounds in sewage to organic acids, including acetic acid, $CH_3COOH$, and propionic acid, $CH_3CH_2COOH$.
2. to result in an effective sludge digestion, acid formers must be equal to the methanogenic bacteria. The main acid former are the obligate anaerobes, which are 10 to 100 times more numerous in digesting sludge than the facultative anaerobes.

- **acid forming fertiliser**
the fertilisers that help in increasing the residual acidity of soil.

- **acid humus**
the presence of humus in the soil that lacks in calcium and

other bases.

- **acid mine drainage**

when certain mines of water drains contain minerals and have a low pH. The low pH is commonly caused by oxidation of iron sulphide to sulphuric acid. Mine water usually contains a high concentration of iron.

- **acid phosphate of lime**

the term refers to the concentrated superphosphate.

- **acid plant**

a plant that is acidic in nature, (pH less than 7). It is due to the production of ammonium salts of organic acids.
For example, malic or oxalic acids.

- **acid scarification**

the process that is used for the modification of hard or impermeable seed coverings. The seed are usually soaked in concentrated sulphuric acid for the purpose of seed scarification. Dry seeds are placed in earthenware containers and covered with concentrated sulphuric acid (S G 1.84) in a ratio of about one part seed to two parts acid. For it, the desirable temperature ranges from 15-27°C . The time of treatment may vary from ten minutes to six hours depending upon the species.

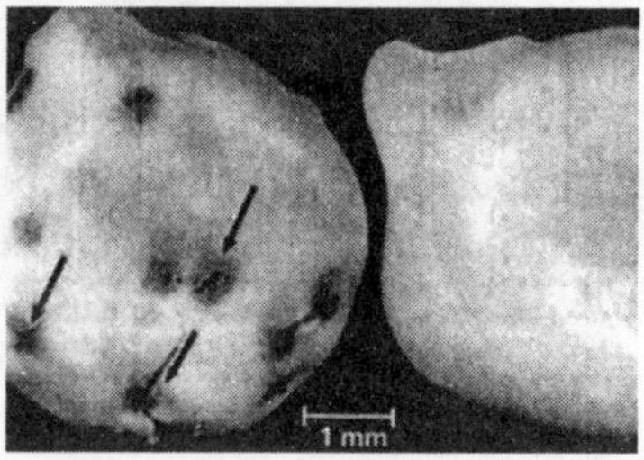


- **acid soil**

refers to the soil with a pH value less than 7.0 or for practical purposes. It is a soil with preponderance of hydrogen ions, and probably of aluminium in proportion to hydroxyl ions. Generally, the term is used for the surface layer or to the root zone unless specified otherwise.

- **acidic amino acid**

amino acids have a net negative charge whereas the pH value is neutral.

- **acidic titrant**

an acid solution used to determine the basicity of another

solution by titration of known concentration.

- **acidifying**
a substance that has the property of lowering the pH.

- **acidity active**
the activity of hydrogen ion in the aqueous phase of a soil. Its activity is measured and expressed as a pH value.

- **acidity potential**
in a soil, the amount of exchangeable hydrogen ion that can be made free or active in the soil solution by cation exchange. It is usually expressed in milliequivalents per unit mass of soil.

- **acidoid**
it refers to a weak acid that is insoluble in water and only superficially dissociated. The colloids are acidoid, capable of retaining and exchangeable cations and H ions on the surface. For e.g., clay.

- **acidometer**
the chemical or a device that is used for the measurement of the amount of acid present in a solution.

- **acidophilic bacteria**
the term refers to the bacteria that can survive at pH as low as 3.0-4.5. It is more properly called aciduric.

- **aciduric bacteria**
bacteria that can grow in medium of a low pH value. For example, *lactobacilli*.

- **acquired characteristic**
due to the interaction of the external and internal environment, a particular characteristic acquired during lifetime.

- **acquired immunity**
immunity that a plant obtains after a period of natural susceptibility.

- **acquired profile characteristics**
due to the process of synthesis, eluviations and alleviation, the characteristics acquired locally.

- **acre**
is a unit of measurement of area:
1 Acre = 4840 Square yards
1 Acre = 0.4047 Hectares
2.4711 Acres = 1 Hectare
640 Acres = 1 Square mile
247.11 Acres = 1 Square km.

- **acre foot**
the volume of water, 43,560 cubic feet, that will cover an area of one acre to a depth of one foot.

- **acre inch**
the amount of water required to cover one twelfth of an

acre-foot, equal to 3,630 cubic feet.

- **acre inch day**
a term principally used to denote one twelfth of an acre-foot, equal to 3,630 cubic feet that is covered in a 24-hr period or •.042 cu ft per sec.

- **acrocentric**
having the centromere located near one end of the chromosome so that one chromosomal arm is long and the other is short.

- **acrodromous**
the veins in a leaf that runs parallel to the edge of the leaf and merge into the tip of the leaf.

- **actinides**
any of a series of chemically similar radioactive elements, with atomic numbers ranging from 89 (actinium) through 103 (lawrencium).

- **actinobiology**
the study of the effects of radiation upon living organisms

- **actinometers**
1. any of several radiometric instruments, such as a pyrheliometer, used chiefly for meteorological measurements of terrestrial and solar radiation.
2. The method is based on the knowledge that every photochemical reaction is sensitive to a particular part of the spectrum in a manner characteristic of the reaction. By observing the extent or rate of reaction, one can determine the photochemical efficiency of different parts of the spectrum.
3. The procedure can be applied also in finding the ratio of light intensities from different sources having the same spectral composition.

- **actinomycetes**
any of an order (Actinomycetales) of filamentous or rod-shaped bacteria (as the actinomyces and streptomyces). They can perform all those functions that are performed by any of the other soil microbes.

- **action**
the causation of change by the exertion of power or a natural process Units of action is the erg-second and the joule-second. Planck's constant is a constant of action.

- **action spectrum**
when the relative rates of photosynthesis are graphically represented in different wavelengths of light.

- **activated carbon**

1. a form of powdered carbon used to absorb recalcitrant organic compounds from waste waters as well as tastes and odours from drinking water.
2. it is also used as an absorbent form of carbon, used to remove odours and toxic substances from gaseous emissions. In advanced waste treatment, activated carbon is used to remove dissolved organic matter from waste water.

- **activated diffusion**

when the solutes move through phloem at a faster rate as a result of accelerated diffusion due to reduced resistance.

- **activated sewage products**

the use of products that have been made from sewage aerated and inoculated with microorganisms. The resulting flocculated organic matter is filtered, dried, ground, screened, bagged, and sold as a fertiliser.

- **activated sludge**

1. material consisting largely of bacteria and protozoa, used in, and produced by, one method of sewage disposal.
2. sewage is mixed with some activated sludge and agitated with air. The organisms of the sludge multiply and purify the sewage, and when it is allowed to settle, they separate out as a greatly increased amount of activated sludge. Part of this is added to new sewage and part disposed off.

- **activated sludge process**

a method of sewage purification in which a little ripe sewage is added to the fresh sewage to be treated, which is then submitted to extensive aeration.

- **activation**

1. when a substance is made artificially radioactive by subjecting it to bombardment by neutrons or other particles.
2. a method for analysing a material for its component chemical elements by bombarding it with nuclear particles or gamma rays and identifying the resultant radiations.

- **activation by light**

during photosynthesis, the change of a complex chemical into another with the help of light energy.

- **activation energy**

the energy that must be provided to the reactants in a chemical reaction to reach an intermediate or activated state from which the products of the reaction can form.

- **activators**

1. a substance that appears to be necessary or beneficial for the action of an enzyme, such as the effect of magnesium and manganese on various phases of alcoholic and lactic acid fermentation.
2. a substance that enhances the ability of a flux to remove oxides and other contaminants from surfaces being joined.

- **active absorption**

when energy is released in contrast to imbibitions and osmosis as the result of absorption of water and other substances against the concentration gradient.

- **active acidity**

the formation of the acid in the solution phase. It is measured by the common procedures as H ion concentration and expressed as pH.

- **active agent**

that part of the pesticide formula that is toxic in nature.

- **active calcium carbonate**

the amount of active calcium carbonate found per 100 grams of fine earth.

- **active element**

the nutrients found in the earth that can be actively used by a plant.

- **active humus**

when the humus present in the soil becomes mineralised because of the presence of the minerals in the soil.

- **active immunity**

immunity naturally produced by the body's own immune system in response to stimulation by foreign antigens. Contrast with passive immunity.

- **active ingredient**

1. in any pesticide product, the component that kills, or otherwise controls, target pests. Pesticides are regulated primarily on the basis of active ingredients.
2. the material or chemical in a batch or formula responsible for producing the desired result and giving the product its main characteristic. Other substances enhance or buffer the active ingredient, and may be classified as inert.

- **active layer**

the top layer of ground subject to annual thawing and

freezing in areas underlain by permafrost.

■ **active site**
the area of an enzyme that accepts a substrate and catalyses its reaction with another. It consists of amoin acids arranged in a configuration specific to a particular substrate or type of substrate. Binding of an inhibiting compound elsewhere on the enzyme molecule may change this configuration and hence the efficiency of the enzyme activity.

■ **active solute absorption**
when energy is released as the result of the intake of dissolved materials by cells against a concentration gradient .

■ **active surface of particles**
a surface which is accessible to the ions circulating in the surrounding electrolyte. It can be the external surface alone or the external surface plus the interlayer surface.

■ **active transport**
an energy expending mechanism by which a cell moves a chemical across the cell membrane from a point of lower concentration to a point of higher concentration, against the diffusion (concentration) gradient.

■ **activity**
1. a measure of the number of disintegrations of a radioactive nuclide per unit of time, reported in units of curies or becquerels.
2. The number of nuclear transitions occurring in a given quantity of radioactive material per unit of time.
3. The capacity of a substance to take part in a chemical reaction.
4. The activity and the concentration of a solute are identical only in very dilute solutions but not otherwise.

■ **activity rate (participation rate)**
the proportion of the total population that is in the working population. The working population is the population that could be

working.

- **actual evapotranspiration (ae)**
the quantity of water lost during the evaporation and transpiration by a crop- soil complex

- **actual lapse rate (climatology)**
indicates temperature values above a given location (in the atmosphere) at a given time. It does not necessarily show a decrease with altitude.

- **acuminate**
the shape of a tip or base of a leaf where the part tapers gradually and often in a concave manner, to a point.

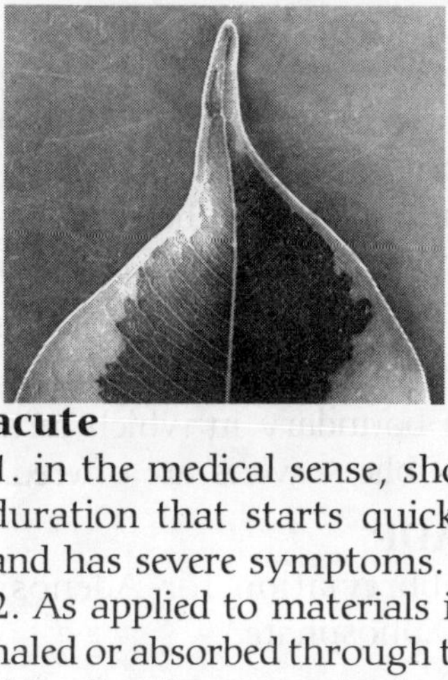

- **acute**
1. in the medical sense, short duration that starts quickly and has severe symptoms.
2. As applied to materials inhaled or absorbed through the skin, it refers to a single exposure. When applied to materials ingested, it refers to a single dose.

- **adaptation**
the adjustment of an organism or population to a new or altered environment through genetic changes brought about by natural selection. The adjustment is physicochemical, enzymic, cytoplasmic, nucleic, morphological and antigenic.

- **adaptive radiation**
the adaptation, the development of many different forms from an originally homogeneous group of organisms as they fill different ecological niches.

- **adaptor compound**
a carrier compound, responsible for the transfer of a substance to the site of reaction. For e.g., RNA

- **additive silage**
any material added to forage at the time of ensiling.

- **adenosine diphosphate (ADP)**
ADP is formed when ATP is broken down within the bodies cell furnace, (the mitochondria). This provides energy for muscular contraction.

- **adenosine monophosphate (amp)**
a nucleotide found in muscle

cells and important in metabolism, reversibly convertible to ADP and ATP.

- **adenosine triphosphate (ATP)**
an organic molecule which stores energy in the cell. It is the last step in the transfer of food energy into the physical work of the muscle.

- **adherence**
the property of sticking together (as of glue and wood) or the joining of surfaces of different composition.

- **adherence of a soil**
the capacity of a soil to adhere to foreign objects, particularly to the blades of farm implements.

- **adiabatic**
without loss or gain of heat to a system. An adiabatic change is a change in volume and pressure of a parcel of gas without an exchange of heat between the parcel and its surroundings.

- **adiabatic colorimeter**
the instrument used for the measurement of the minimum loss of heat in a chemical reaction.

- **adiabatic process**
a thermodynamic change of state of a system such that no heat or mass is transferred across the boundaries of the system. In an adiabatic process, expansion always results in cooling, and compression in warming.

- **adjuvants**
materials which act as wetting or spreading agents, stickers, penetrants or emulsifiers and help many pesticides to work more efficiently.

- **adlittoral**
region which is dry most of the year and is submerged only during very high tide, an excess of rain or a very high sea. Its vegetation is largely composed of halophytes.

- **administered price**
a price that is not determined by competitive market forces, but by the conscious act of an administrative body, such as the government, a marketing board, or a trading group.

- **administrative area**
a boundary in which certain public services are offered.

- **ADP**
abbreviation for Adenosine Diphosphate.

- **adsorbed**
being accumulated on a surface of a solid with an absorbed material being taken

inside and probably held more tightly than the adsorbed material.

- **adsorbed cations**

this refers to cations which are held by small, negatively charged particles of clay and organic matter called micelles. Micelles can be considered as polyanions because they carry large numbers of negative charges.

- **adsorption**

1. the adhesion of the molecules of gases, dissolved substances, or liquids in more or less concentrated form, to the surface of solids or liquids with which they are in contact.
2. Commercial adsorbent materials have enormous internal surfaces. It is often used to extract pollutants by causing them to be attached to such adsorbents as activated carbon or silica gel. Hydrophobic, or water-repulsing, adsorbents are used to extract oil from waterways in oil spills. The adherence of a gas, liquid, or dissolved material on the surface of a solid.

- **adsorption chromatography**

the separation of a mixture of adsorbed substances by charge or other property by allowing the partition of the mixture between a moving phase and a stationary phase.

- **adsorption coefficient (equilibrium coefficient)**

the capacity of soil to adsorb a herbicide.

- **adsorption complex**

the group of substances in soil which are capable of adsorbing materials, the organic matter and colloidal clay forming the greater part of the complex and materials in silt and sand-size exhibiting adsorption, but to a reduced extent in most soils.

- **advanced waste treatment**

the process of waste water treatment beyond the secondary or biologic stage, it includes removal of such nutrients as phosphorus and nitrogen and a high percentage of suspended solids. Advanced waste treatment, known as tertiary treatment, is the 'polishing state' of waste water treatment and produces a high-quality effluent.

- **advection**

Process of transport of an atmospheric property, or substance within the atmosphere, solely by the mass motion of the atmosphere. In air, advection may result in due to the transfer of heat and of water vapour from warm sea to a cooler land and vice-versa.

- **advection fog**
a fog that forms when warm air flows over a cold surface and cools from below until saturation is reached, but is most frequently caused by warm moist air from the sea, moving over the land in winter after a cold spell.

- **advection frost**
a kind of frost which results when the temperature at the surface in an air mass is below freezing point. It is also known as air mass frost.

- **advective freeze**
a large cold air mass usually accompanied by windy conditions and therefore no temperature inversion occurs.

- **adventitious**
1. organ or structure in an unusual place such as roots growing from the stem.
2. a disability that develops at any time after birth.

- **adventitious bud**
a bud produced irregularly on older portions of a plant and not at the tip of the stem or in the leaf axil, not connected with the pith of a tree. Usually such buds are in response to cambium injury.

- **adventive**
term to indicate that a plant has become naturalised or temporarily established in an area.

- **aeolian soil**
a type of soil composed of windblown materials, wind transported deposits. For e.g., sand dunes, wind deposits of fine volcanic ash.

- **aerate**
to add air to water or to the pores in soil. It adds oxygen and allows other gases trapped in the water to escape.

- **aeration**
the process by which air in the soil is replaced by air from the atmosphere. The rate of aeration depends largely upon the volume of pores in the soil.

- **aeration factor**
the amount of oxygen consumed by soil microbes under optimum respiratory condition to the consumption in the presence of soil expressed in percentage of the optimum value. The limitation of this index is due to its artificial condition, which does not represent true natural condition of soil.

▪ **aeration porosity**
the pore space filled with air when the soil is under a tension of a 50-cm column of water. This value is also called aeration capacity or noncapillary porosity. It corresponds to a pore size of 0.06 mm in diameter or larger. It is used as a quantitative standard by which the potential air capacities of soils can be compared.

▪ **aeration soil**
1. the process by which air and other gases in the soil are renewed. The rate of soil aeration depends largely on the size and number of soil pores and on the amount of water clogging the pores.
2. a soil with many large pores open to permit rapid aeration is said to be well aerated, whereas a poorly aerated soil either has few large pores or most of those are blocked by water. In a well-aerated soil, the soil air is very similar in composition to the atmosphere above the soil. Poorly aerated soils usually contain a much higher percentage of carbon dioxide and a correspondingly lower percentage of oxygen than the atmosphere above the soil.

▪ **aeration tank**
a tank for mixing and aeration of sewage and activated sludge—normally 3 to 5 m deep. It is the first stage of secondary treatment of wastewater. Here, sewage is mixed with air and bacteria so that any organic material is broken down.

▪ **aerator**
mechanism, which mixes air into water, soil, sewage, etc. Alternately, the part on the end of a sink spout which mixes air with the water to reduce splattering. Four types exist—cascade or free-fall aerators, spray aerators, injection aerators and surface aerators. All types create useful turbulence at the air-water interface.

▪ **aerial**
of or pertaining to the air, atmosphere, or aviation.

- **aerial application**
treatment applied with the use of an aeroplane or helicopter.

- **aerial dusting**
dusting of pesticide, fungicide and herbicide powders from an aeroplane.

- **aerial irrigation**
also called overhead or sprinkler irrigation. The conveying of water under pressure through pipes to the area to be irrigated and passed out through nozzles of sprinklers.

- **aerial photograph**
a photograph of the earth's surface taken from airborne equipment, sometimes called aerial photo or air photograph.

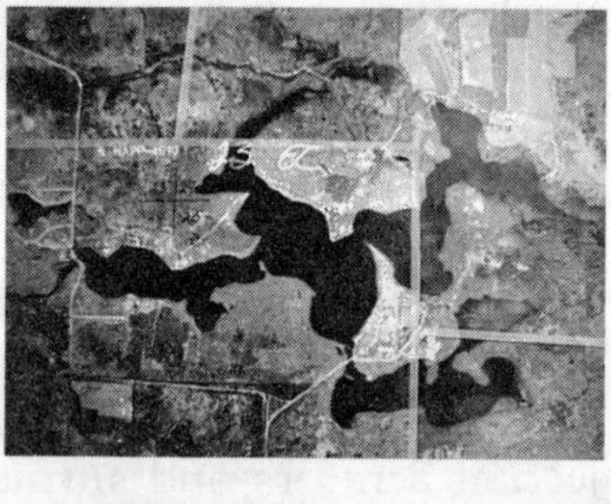

- **aerial root**
a root, usually arising adventitiously from a stem. Epiphytes produce aerial roots which have specialised tissue called velamen for absorbing moisture from the air. Aerial roots may contain chlorophyll and photosynthesise.

- **aerobic**
1. characterising organisms able to live only in the presence of air or free oxygen, and conditions that exist only in the presence of air or free oxygen.
2. process requiring oxygen.

- **aerobic bacteria**
bacteria that requires the presence of free or dissolved oxygen in their environment for survival and reproduction.

- **aerobic digester**
an aeration tank that treats waste activated humus or primary sludges or a mixture of them, usually in a small plant with extended aeration or contact stabilisation treatment. The high cost of aeration and the absence of fuel gas are disadvantages compared with anaerobic sludge digestion, but the reduction of volatile solids is about equal and the supernatant liquid has a lower BOD. In addition, the sludge has no smell, is stable and easily de-watered, there are fewer operational problems and the detention period is shorter than anaerobic digestion.

- **aerobiosis**
life sustained in the presence of air or oxygen.

- **aerography**
the subject that seeks to describe the properties, dimensions etc., of the atmosphere.

- **aeropedoturbation**
the mixing of soil by movement of gases in the soil, during and after rains.

- **aerosol spray**
a means of atomising liquids by propelling from a pressurised container by a soluble gas.

- **aerosols**
a colloidal system in which the dispersion medium is a gas.

- **aerotolerant bacteria**
colloids of liquid or solid particles less than 10 microns in diameter suspended in gas.

- **affination**
the process of removing molasses film from raw sugar by mingling with thin syrup or water and fugalling.

- **affinity**
attractive force between substances or particles causing them to combine chemically.

- **affluent**
a branch that flows into the main stream.

- **afforest to**
to establish a forest on previously unforested land.

- **afforestation**
establishing a forest on an area which has not previously had trees growing on it.

- **after ripening**
the collective name for processes which are necessary before germination can take place in certain seeds, even though external conditions may be suitable. A period of dormancy is thus imposed on the seed, preventing premature germination before an unfavourable season such as winter.

- **aftermath (rowen)**
when the meadow plants are grown in the same season after hay or seed crop has been cut.

- **ag soil**
denotes to a soil that has a black A horizon over gleified

Q horizon due to high water table.

- **agalol (ceresan-universal nassberge, aretan, met-hoxy ethyl mercuric chloride)**
is a mercurial seed disinfectant. It is sold as a formulation containing 1.75% mercury.

- **agar**
mixture of polysaccharides derived from red algae that forms a gel at temperatures below about 40°C. Used as a support medium, when supplemented by appropriate buffers and/or nutrients and other ingredients, for the production of microbial cultures, overlaying tissue culture cells, electrophoresis, etc.

- **agglomerate soil**
a soil with the grains uncoated, or with a loose friable coat. The non-mineral material forms aggregates in the spaces between the mineral particles. The particles are not stuck together, or only slightly so.

- **agglomeratic structure**
a kind of microstructure in which mineral grains are completely bare, forming only easily shifted, flocculated or insoluble granular deposits.

- **agglusol**
a synthetic resin used to restore a structure to the soil.

- **aggradation**
accumulation of sediment in a stream channel on an alluvial fan or on a floodplain.

- **aggregate**
1. to bring together, to collect or unite into a mass.
2. composed of a mixture of substances separable by mechanical means.
3. the mineral materials, such as sand or gravel, used in making concrete. Fine aggregate may be considered as the material that passes a 6.35 mm screen, and coarse aggregate as the material that does not pass a 6.35 mm screen.

- **aggregate analysis**
a soil analysis which measures the relative distribution of the various sizes of aggregates and permits a calculation of the percentage aggregation of the finer mechanical separates. It provides an index to soil structure, but does not characterise the type of structure.

- **aggregate formation**
is primarily the orientation of the finer soil particles and bringing them so close together that the physical forces

hold them firmly when allowed to dry.

- **aggregate stabilisation**
is concerned with both the forces and conditions that exist within the aggregate, and with those that exist on the outside. The former are so much important that the latter are often ignored.

- **agmark**
1. certain standard grades fixed by the Marketing Department of Government of India. Products conforming to these standards can be stamped with the designation agmark and sold under that title. Agmark title is well recognised by the trade.
2. Standards have been prescribed for products like fruits, vegetables, eggs, dairy produce (ghee), tobacco, coffee, skins, cotton and rice.

- **agon**
a kind of plough which opens a furrow deeper than the raise to improve a terrain which is swampy.

- **agrad**
a cultivated plant.

- **agrarian legislation**
acts enacted from time to time to prevent a farmer from being reduced to the position of a tenant or an agricultural labourer by private moneylenders and to prevent absentee landlordism.

- **agrarian system**
the pattern of land ownership and distribution, social and institutional structure of the agricultural economy.

- **agrarianism**
a theory or practice based on the idea that landed property should be evenly or fairly distributed among the people of a nation and that legislation should be enacted to foster the more general occupation and use of the land.

- **agreement water rate**
a charge for the supply of irrigation from the Government canals. This is a fixed rate for a number of years whether the supply of water is utilised or not in that period.

- **agrestal**
the process of growing in cultivated land, but not itself cultivated, for e.g., a weed.

- **agric horizon**
anilluvial horizon of day and organic matter accumulation just under the plough layer resulting from long-continued cultivation.

- **agricultural commodity**
  any plant or part of a plant, animal, or animal product that is to be bought or sold.

- **agricultural credit**
  the amount of money needed by a farmer to achieve a proper combination of productive factors like land labour, inputs, livestock, machinery and marginal ability so that the planned level of income is generated by his farm.

- **agricultural economics**
  a branch of agricultural science which uses the principles and techniques developed in agronomy, farm business, animal husbandry and economics in a coordinated way for the economic benefits and welfare of a farmer in particular and a region or state or country or the world in general.

- **agricultural engineering**
  the branch of engineering dealing with physical problems involved in the preparation of land and the production, processing and storage of agricultural products.

- **agricultural extension services**
  refers to the services normally offered by the Government to farmers in various fields in order to help them to raise productivity.

- **agricultural geology**
  the branch of applied geology, which deals with the nature and properties of rocks and minerals of importance in soil formation.

- **agricultural liming material**
  a material whose calcium and magnesium content is capable of neutralising soil acidity.

- **agricultural marketing**
  a branch of marketing which covers the services or functions, which take part in bringing product from a farmer to the final consumer.

- **agricultural mechanisation**
  the introduction of machinery in the agricultural production process.

- **agricultural operations**
  there are various agricultural operations, common among them are: ploughing, clod crushing: levelling, disking,

harrowing, manuring, compacting the soil, seed bed preparation, comprise harrowing, preparing ridges and furrows, basins, raised beds, etc, sowing, gap filling, thinning, weeding, mulching, top dressing: irrigation, harvesting, threshing and winnowing.

- **agricultural planning**
  a system of organisation for the analysis and adjustment of state and local agricultural problems.

- **agricultural pollution**
  Liquid and solid wastes from any types of agricultural enterprise, including chemical runoff from farm fields, and feedlots, erosion and dust from ploughing, animal manure and carcasses, and debris.

- **agricultural region**
  an extensive area of land with broadly similar conditions and patterns of agricultural practice.

- **agriculture**
  the raising of crops and animals for food, feed, fibre, fuel, or other useful products. To some extent, it deals with preparation of plant and animal product for human consumption and marketing of these products.

- **agrisilviculture**
  refers to the combination of forestry and agricultural crop production either simultaneously or in rotation.

- **agrium**
  a formation on cultivated land.

- **agrobacterium**
  a common soil bacterium that can naturally carry genetic information (DNA) into plant cells. Often used in genetic modification.

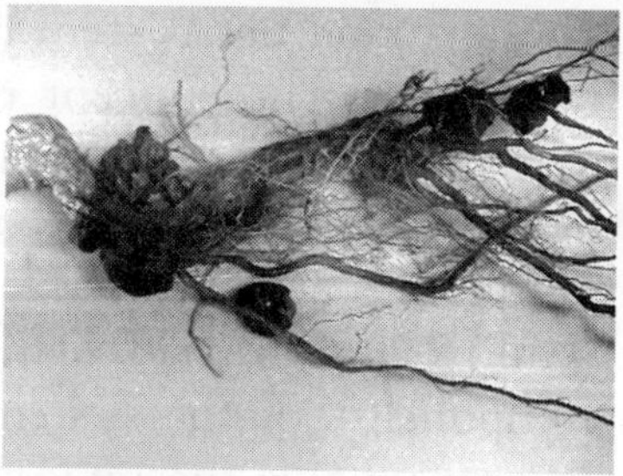

- **agrobiology**
  the study of plant nutrition and growth especially as a way to increase crop yield.

- **agroceric acid ($c_{26}$ $ho_3$)**
  a constituent of humus.

- **agroclimatology**
  the study of those aspects of climate, which are relevant to the problems of agriculture.

- **agroecology**
  the study that shows the relation of crop varietal characteristics to their adaptation to environmental conditions.

- **agroforestry**
  1. a collective term for systems of land use in which woody plants (trees and shrubs) are deliberately combined on the same land management unit with herbaceous crops and/or animals, either in some form of spatial arrangement or in sequence.
  2. For a land use system to fall within the concept of agro-forestry, these should be both ecological and economic interactions between woody plants arid other components of the system. It may involve the integration of trees into farming systems or crops and livestock into forests. In practice, a high proportion of agro-forestry systems involve the growing of trees on what is primarily agricultural land.

- **agrological map**
  a map giving the agricultural properties of soils.

- **agrology**
  the study of the properties of the soil in relation to crops. It differs from pedology in that the latter studies the soil for itself.

- **agronomy**
  branch of agriculture dealing with field-crop production and soil management. The name is derived from the Greek words *agros* (field) and *nomos* (to manage). Thus, it is a branch of agriculture dealing with the growing of field crops like cereals, millets and oil seeds. It lays emphasis on production of crops, including experimental work aimed at increasing yields, controlling diseases, developing special varieties, improving cultivation and selecting for suitability to environmental conditions. Soil enrichment and conservation are also among its important objectives.

- **agrosterine**
an intermediate product in the decomposition of organic matter in the soil, having a formula $C_{24}H44O$ or $C_{26}H_{4}40$.

- **agrosterol ($e_{26}h_{44}o$)**
a constituent of humus.

- **agrostology**
the study of grasses, their classification, management and utilisation.

- **agrotechnical structure (soil)**
the structure produced by the burrowing and digging animals inhabiting the soil. This structure is formed on ploughed soils by agricultural implements. Its characteristic features are the roundness of aggregates and the diffuseness of faces and edges. Its stability depends on the level of agricultural techniques. This structure may be encountered in the arable horizons of all soil zones.

- **ai horizon**
a horizon made up of minerals, formed or forming or at adjacent to the surface, in which the feature emphasised is an accumulation of humified organic matter intimately associated with the mineral fraction.

- **aihg**
active ingredient per hundred gallons.

- **air borne dust**
dust larger than 10 microns settles quickly in air. At 1 micron size, however, dust settles so slowly (about 0.03 mm/s, varying with density) that it can be considered airborne at and below this size.

- **air capacity of soils**
when air is retained in the soil after being saturated with water and allowed to drain for 24 hours *i.e.* the soil at field capacity moisture, it is called air capacity of soil. Alternatively, the air capacity is equal to the non-capillary pore space of soils. It is affected by texture, structure, organic matter, drainage, tillage, compaction cracking, root penetration etc. all of which influence the non-capillary pore space of soils.

- **air content (soil)**
the ratio of the volume of air to the total pore space when measured.

- **air dry soil**
the stage of dryness at equilibrium with the moisture content in the surrounding atmosphere. The actualmoisture con-

tent will depend upon the relative humidity and the temperature of the surrounding atmosphere!.

- **air entry value (tensiometer cup)**
  it refers to the limit of maximum tension up to which the cup is functional. Beyond this value, the meniscus of water in the capillaries of cup ruptures and air enters into the system and the instrument ceases to function. A well-prepared cup can have the air entry values as high as O.v bar.

- **air filled porosity**
  the ratio of pores filled with air or pores free from water to the total porosity. This is the measure of the relative air content of the soil.

- **air meter**
  refers to an anemometer consisting of wheel with a calibrated deal which rotates in the wind.

- **air monitor**
  a detecting device, used for control and warning purposes, to measure the amount of radioactivity present in the air.

- **air pollution**
  the presence in the outdoor atmosphere, of any dust, fumes, mist, smoke, other particulate matter, vapour, gas, odorous substances, or a combination thereof, in sufficient quantities and of such characteristics and duration as to be, or likely to be, injurious to health or welfare, animal or plant life, or property, or as to interfere with the enjoyment of life or property.

- **air porosity**
  it refers to the that portion of the soil that is filled with air at any given time or under a given set of conditions.

- **air pycnometer**
  a standard vessel used in measuring the density or specific gravity of air.

- **air slaked lime**
  a product composed of varying proportions of the oxide, hydroxide, and carbonate of calcium, or of calcium and magnesium, and derived from

the exposure of quicklime to the weather.

- **air space ratio**
ratio of:
1. volume of water that can be drained from a saturated soil under the action of force of gravity.
2. total volume of voids.

- **air vents**
vertical pipe structures to release air entrapped in the pipeline and to prevent vacuum. Entrapped air must be removed to permit an even flow and avoid the danger of water hammer.

- **air void ratio**
ratio of:
1. the volume of air space.
2. the total volume of voids in a soil mass.

- **airlift pump**
1. a pump with no moving parts, consequently most useful for lifting abrasive material such as gritty water in sewage works.
2. Two pipes of different diameter hang in the well. The smaller is sometimes within the larger. Air is blown down the smaller pipe and from it passes up the larger one. Air bubbles and water rise together in the larger pipe. Large volumes of very dirty fluid can be pumped though at high expense in compressed air and at low mechanical efficiency. For pumping sewage the airlift pump has the advantage of aeration of the water, reducing smells.

- **alaun ($t_7$)**
soil characterised by percolates of iron and alumina sulphates in very fine dispersed form. The soil filter is acid.

- **albert's rainfall factor**
the quotient of the precipitation divided by the average temperature for the period without frost. It is the Lang factor, corrected.
If P=annual rainfall
p (mm), T=average temperature
If F=40-60 laterisation is possible

- **albic horizon**
a horizon from which clay and free iron oxides have been removed. Generally equivalent to Aa horizon.

- **albino**
1. a more or less white plant.
2. An abnormal form lacking pigment.

**albus**
a Latin term that means white.

**alcohol thermometer**
thermometer consisting of a glass capillary tube marked with degrees Celsius or Fahrenheit and containing alcohol which rises or falls as it expands or contracts with changes in temperature.

**alfalfa**
a perennial legume rich in vitamins, minerals, and protein. When harvested before blooming, alfalfa is a high energy and low fibre source of food for herbivores.

**alfalfa gate**
in irrigation, a light sheet-metal slide gate in a section of pipe, for the control of water from sublaterals into fields and ditches.

**alfalfa valve**
a valve consisting of a horizontal disc equal in diameter to the riser pipe and located at the top of the riser pipe in an underground piped irrigation water supply system. There is a ring around the base outside of the disc to provide a seat and seal for a portable hydrant.

**alfisols**
an order of soils with a medium-to-high base supply, horizons of clay accumulation, and gray-brown surface horizons. These soils have formed mostly under forest or savannah vegetation in climates with slight to pronounced seasonal moisture deficit.

**algae**
microscopic plants which contain chlorophyll and live floating or suspended in water. They also may be attached to structures, rocks or other submerged surfaces. They are food for fish and small aquatic animals.

**algaecide**
A chemical agent specifically designed and used to kill algae and prevent their recurrence. Algaecides are often applied to water to control nuisance algal blooms.

**alien**
In a natural system, a plant or animal not evolved in that system and to which the system is not adapted.

**alkali flat**
an alkaline, marshy area in an arid region in which one or

more direct streams lead. It becomes a barren area of hard mud covered with alkali when all the water is evaporated in the dry season, after heavy rainfall, it becomes a shallow muddy lake.

- **alkali fusion**
calcining in the presence of potassium and sodium carbonate a method for quantitative analysis of the mineral colloids of the soil.

- **alkali health**
*Frankenia grandifolia* var. *Campestris* is an indicator plant, has luxuriant growth of uniform stand usually on highly saline soil unsuited for agriculture unless drained and leached. Sparse growth indicates less salt.

- **alkali sacation**
*Sporobolus airoides* is an indicator plant having pure vigorous stand on wet, saline, or saline-alkali soils with higher water table. Drainage and leaching required and often gypsum amendment.

- **alkali soil**
soil containing a higher concentration of mineral salt than natural acid. Alkyd Thermoplastic or synthetic resin used in making paint and other coatings. Used in both interior and exterior paint, it is a thinned solvents.

- **alkaline water**
when there is high per cent sodium (approaching and exceeding 60) in water, but it is relatively low in total dissolved solids.

- **alkalisation**
a process whereby the exchangeable sodium content of the soil is increased.

- **alkaloid**
one of a group of organic alkaline (basic) substances obtained from plants. It reacts with acids to form salts. They contain nitrogen and usually occur as salts of acids such as malic, citric and succinic acids. Their function in plants remains obscure, but it is suggested that they may be nitrogenous end-products of metabolism, or they may have a protective function against herbivores. Important examples are quinine, opium, morphine, nicotine, atropine, codenine.

- **allergy**
an acquired, abnormal immune response to a substance that can cause a broad range of inflammatory reactions.

- **allitic soil**
a soil in which the clay fraction has a high proportion of

aluminium and iron compounds, and from which silica has been removed.

- **available factor**

1. the ratio of the time.
2. a piece of equipment or a unit, is ready for or in service to the total time interval under consideration.

- **average**

1. a single number typifying or representing a set of values of which it is a function.
2. a single number computed such that its value is not less than the smallest nor larger than the largest of a set values.

- **average cost (ac)**

the total cost of production divided by the total quantity produced. Under conditions of perfect competition or monopoly, it is usual to include normal profit while calculating the total cost of the firm.

- **average deviation**

1. the average deviation is one of several indices of variability that statisticians use to characterise the dispersion among the measures in a given population.
2. To calculate the average deviation of a set of scores it is first necessary to compute their mean and then specify the distance between each score and that mean without regard to whether the score is above or below the mean. The average deviation is defined as the mean of these absolute values.

- **average error**

the mean of all the errors taken without regard to sign.

- **average fitness of a population**

the average fitness of the genotypes in a population weighted by their frequencies.

- **average holding**

figures arrived at after dividing the total area of holding by the total number of agricultural holdings.

- **average propensity to consume (APC)**

total consumption in a given period divided by total disposable income.

- **average sample number**

a term used in sequential analysis to describe the mean number of observations that must be made before making a decision to accept or reject a hypothesis with prestated risks of alpha and beta error.

- **avogadro's law**
  states that equal volumes of all gases under the same conditions of pressure and temperature contain the same number of molecules.

- **avogadro's number**
  the number of particles in a mole of these particles: 6.0220 x 1023.

- **awn**
  a slender, bristlelike appendage found on the spikelets of many grasses.

- **axis of earth**
  the diameter between the north and south poles, about which the earth rotates in 24 hours. The length of the axis is 12714 Km.

- **azonal soil**
  soil without distinct genetic horizons. Such soils may have non-Chernozemic Ah horizons, or thin Chernozemic Ah horizons, but lack B horizons.

- **azotobacter**
  any of various rod-shaped, nonpathogenic, nitrogen-fixing bacteria of the genus *Azotobacter,* found in soil and water.

- **azureus**
  the Latin for 'a fine blue colour'.

- **B**
  this symbol is added to the designation of a buried genetic horizon or horizons. Horizons of another solum may or may not have formed in the overlying material, which may be similar to, or different from, the assumed parent material of the buried soil.

- **b horizon**
  the soil layer just below the A horizon that has about the same colour and texture throughout its depth. It is usually 10 to 12 inches thick but may range from 4 inches to 8 feet.

- **$B^+$**
  negative beta particles are sometimes called negatrons B. The term beta particle and the, symbol B are reserved for electrons originating in a nucleus.

- **bacillus amylobacter**
  one of the most important among bacilli which transform the cellulose of the plant tissues in litter into humus.

- **bacillus mycoides**
  the bacteria which reduce nitrates to nitrites.

- **background**
  in measurements of radioactivity, the observed count, in the absence of a sample caused

by cosmic radiation, instrument noise, power line fluctuations, etc.

- **backing**
counterclockwise shift in the wind direction in the Northern Hemisphere at a certain location. In the Southern Hemisphere, it is clockwise. This can either happen in the horizontal or the vertical (with height).

- **backscattering**
small fraction of light that is reflected out of the original direction of propagation by scattering suffers a reversal of direction. In other words, it propagates in the optical waveguide towards the transmitter.

- **backwash effect**
an unfavourable situation arising in underdeveloped countries because of the over-emphasis on export trade (usually raw materials) at the expense of the growth of domestic manufacturing and industrialisation of rural areas.

- **bacteria**
Single-celled organisms, free-living or parasitic, that break down the wastes and bodies of dead organisms, making their components available for re-use by other organisms.

- **badland**
is steep or very steep nearly barren land, ordinarily not stony, broken by numerous intermittent drainage channels. Badland is most common in semiarid and arid regions, where-streams have entrenched themselves in soft geologic materials. Local relief generally falls between 25 and 500 feet. Runoff is very high, and geological erosion active. Badland has practically no agricultural value, except for small areas of soil with some value for grazing that may be included in the mapping unit. The relief is similar to that of rough broken land, which has a cover of vegetation.

- **bag or tag information**
in the fertiliser marketing each bag of fertiliser sold must bear information either on the bag itself or tagged with the container, showing net weight of fertiliser, brand and grade of fertiliser, guaranteed analysis, and name, address of the manufacturer.

- **bagasse**
is the waste fibre left over from the sugar cane after it

has been processed to make sugar. Bagasse can be burnt and used to create electricity. It is a type of biomass.

- **bairum polysulphide (solbar)**
  it is a fungicide prepared by saturating barium hydroxide with hydrogen sulphide or by mixing ground sulphur with barium sulphate powder.

- **baked clay tile lined channel**
  when irrigation channels are lined with semicircular clay tiles moulded and burnt in kilns and ends are joined with cement mortar.

- **baked soil**
  soil which has a hard, or even glazed, surface due to a wet condition followed by an intense drying action of the sun.

- **balance**
  a beam balance of great precision used in quantitative chemical analysis e.g., *nation* into balance so that reaction substances and reaction products obey the laws of conservation of mass and charge.

- **balance of nature**
  the equilibrium that is kept in a natural community such that the numbers of species are kept fairly constant and don't increase to such an extent as to deplete food sources. Nature is in a constant state of change and this constant evolutionary process is nature's way of trying to create a balance, which is never reached.

- **balanced fertilisation**
  the process of application of fertilisations in such proportions that a balance is maintained in the supply of various plant nutrients. This is done to facilitate maximum efficiency in plant nutrition.

- **balanced fertiliser**
  a soil additive containing suitable proportions of each necessary mineral element to develop a plant or a crop.

- **balanced growth**
  it refers to a programme of coordinated growth of all sectors of the economy. Developing nations plan a balanced growth of their economies.

- **balanced growth path**
a pattern of economic growth that takes place in such a way that demand equal supply in all markets.

- **balanced nutrition through fertilisation**
the continuous availability to plants of all the essential elements, with none in excess and none deficient. Balanced nutrition of plants in relation to animals can also mean applying fertilisers to increase the nutritional value of the forage or feed.

- **balanced ration**
food which provides the various nutrients in the proportions and amounts necessary for the proper nourishment of an animal.

- **balanced solution balanced medium**
a culture or solid medium with the elements necessary for growth, in such proportions that the toxic effect of individual salts are mutually eliminated. They are used for the growth of algae, bacteria, fungi.

- **bale**
a large compressed, bound, and often wrapped bundle of a commodity, such as cotton or hay.

- **ball planting**
method of planting young plants, in a process plants are taken from the ground with a compact ball or lump of earth round their roots to provide nourishment and help their growth in a new location.

- **ballast elements**
elements such as aluminium and silicon have been called ballast elements, because, they are generally present in large amounts, although the plant can grow normally without them.

- **balling**
a measurement of sugar percentage in sample syrup at fixed temperature.

- **band**
a long ridge-like hill.

- **band placement**
placement of fertiliser at the time of sowing in a band adjacent to but not in contact with seed.

- **band seeding**
placing the seeds in rows directly above, but not in contact with a band of fertiliser.

- **band treatment**
to apply an herbicide in bands, usually 12 to 15 inches wide, directly over the crop rows.

- **bandhara**
a prominent source of irrigation in Maharashtra (India). This consists of one or more dams for deflecting water from the watercourse during rainy season and retaining the same for irrigation.

- **banding fertiliser**
when fertiliser is placed in the soil in continuous narrow bands or ribbons, generally at specified distances from the plants or seeds. Usually below the soil surface.

- **bank**
1. the sloping ground that borders a stream and confines the water in the natural channel when the water level, or flow, is normal.
2. the margins of a channel. Banks are called right or left as viewed facing in the direction of the flow.

- **bank erosion**
land or soil losses of the banks of rivers, lakes, ditches or canals resulting from currents, wave cutting, undermining, caving, rilling, or gullying of slopes.

- **bank storage**
the ability of water to flow into the bank of a river at high water and to return to the river at low water, acting as groundwater in the interval. Gravelly or other porous soils have greater bank storage than clays, silts or impervious soils.

- **bar**
a unit of pressure, equal to the sea-level pressure of Earth's atmosphere, 1 bar = 0.987 atmosphere = 101,300 pascals = 14.5 lbs/square inch = 100,000 Newtons per square meter.

- **bar chart**
a chart that graphs the high, low, and settlement prices for a specific trading session over a given period of time.

- **bar graph**
a graphical way of showing quantitative comparisons by using rectangular shapes with lengths proportional to the measure of what is being compared.

- **bar screen**
in wastewater treatment, a

device used to remove large solids.

- **barbier morgan index**
  an index of the chemical fertility of the soil.

- **bargaining theory of wages**
  according to this theory, the wages in an industry are determined by the relative bargaining strength of the trade unions and employers' association concerned.

- **bark**

the protective outer layer of the roots, trunk and branches of a tree or woody plants. Bark may be thick and rough or thin, smooth and papery.

- **barley**
  cereal grain, from which most beer is made. To make beer, the barley must first be malted, then mashed, then fermented.

- **barn**
  1. an outlying farm building for storing grain or animal feed and housing farm animals.
  2. a unit of area for measuring a nuclear cross section. One barn equals 10-24 square centimeter.

- **barogram**
  the continuous record of atmospheric pressure-made by a barograph.

- **barograph**
  an instrument used to keep a record of atmospheric pressure, such as on a paper drum.

- **barometer**
  an instrument, often using a column of mercury, for measuring atmospheric pressure. Traditionally a falling barometer measurement meant an approaching storm.

- **barothermograph**
  an instrument for recording both pressure and temperature, as of the atmosphere.

▪ **barrage**
any artificial obstruction placed in water to increase water level or divert it. Usually the idea is to control peak flow for later release.

▪ **barrel**
a unit of volume measure used for petroleum and refined products. 1 barrel = 42 U.S. gallons.

▪ **barren lands**
1. those ecosystems in which less than one third of the area has vegetation or other cover. 2. in general, barren land has thin soil, sand, or rocks. Barren lands include deserts, dry salt flats, beaches, sand dunes, exposed rock, strip mines, quarries, and gravel pits.

▪ **barrier application**
when pesticide or any other agent is used to stop pests from entering a container, area, field or building.

▪ **barrier lake**
a lake that has formed where rocks have built up to stop water from flooding away.

▪ **barshy soil**
soil composed of coarser angular rock fragments with fine soil.

▪ **basal treatment**
an application to the stems of plants at or just above the level of the soil.

▪ **basalts**
a fine-grained igneous rock dominated by dark-coloured minerals, consisting of plagioclase feldspars (a calcium-sodium-aluminium-silicate, usually present in amounts over 50%), and ferromagnesium silicates. Basalts and andesites represent about 98% of all extrusive igneous rocks.

▪ **base date**
when a date is recorded at the starting point for an index, usually given the value 100.

▪ **base exchange**
the process whereby a soil absorbs certain (positively charged ions) from the soil

solution and gives up other cations in equivalent quantities, the colloidal fraction of the soil is primarily concerned. The term cation exchange is tending to be substituted.

- **base exchange capacity**
the amount of base held on a clay under specified conditions for e.g., of pH.

- **base level**
the lowest level to which a stream can erode the channel through which it flows, generally equal to the prevailing global sea level.

- **base pair**
the basic units of DNA, base pairs are chemical structures made up of the chemicals adenine, thiamine, guanine and cytosine, which are designated by the letters A, T, G and C, respectively.

- **base pairing rules**
the requirement that adenine must always form a base pair with thiamine (or Uralic) and guanine with cytosine in a nucleic acid double helix.

- **base ratio**
the ratio between the amount of adenine and thiamine bases (A+T) and the amount of guanine and cytosine bases (C+G) in a DNA sample.

- **base rich**
soils containing a relatively large amount of free basic ions such as calcium or magnesium.

- **base run off**
sustained or fair weather runoff. In most streams, base runoff is composed largely of groundwater effluent. The term base flow is often used in the same sense as base runoff. However, the distinction is the same as that between streamflow and runoff. When the concept in the terms base flow and base runoff is that of the natural flow in a stream, base runoff is the logical term.

- **base saturation**
the degree to which material having base exchange properties is saturated with exchangeable bases (sum of Ca, Mg, Na, K), expressed as a percentage of the exchange capacity.

- **base saturation ratio**
ratio of the base content to the humus content. The affinity of humus for the bases decreases as the ratio increases.

- **base year**
the benchmark year from which a facility measures

progress in reducing the byproducts and emissions attributable to the production unit. In each succeeding reporting year the byproducts/ emissions attributable to the production unit are compared to those in the base year, on a per unit of product basis.

- **basic crops**
crops which because of acreage, value, climate, etc., are considered important in agricultural economy, paddy, wheat, cotton, sugarcane, etc.

- **basic data**
records of observations and measurements of physical facts, occurrences, and conditions, as they have occurred, excluding therefrom any material or information developed by means of computation or estimates as are necessary to present a clear statement of facts, occurrences, and conditions.

- **basic grassland**
a type of grassland which occurs on chalk and limestone terrain.
For e.g., *Festuca ovina* and *Festuca rubra*.

- **basic number**
the haploid number of chromosomes, usually represented by n or x.

- **basic phosphate slag**
a by-product of the manufacture of steel from phosphatic iron ores. It contains not less than 12 per cent total phosphoric acid of which at least 80 per cent is available (citrate-soluble). It is ground so that not less than 70 per cent of the material passes through a 100-mesh sieve and 90 per cent passes through a 50-mesh sieve.

- **basic slag**
slag produced in making steel, low in silica but having large amounts of calcium phosphate, useful as fertiliser. The impurities in the iron consisting of iron, silica and phosphorus are fluxed with lime and sagged off. The material is cooled, finely ground and bagged. It contains 8-10% $P_2O_5$. Basic slag also contains traces of Zn, Cu and B. It is much valued for its lime content on acid soils.

- **basifixed**

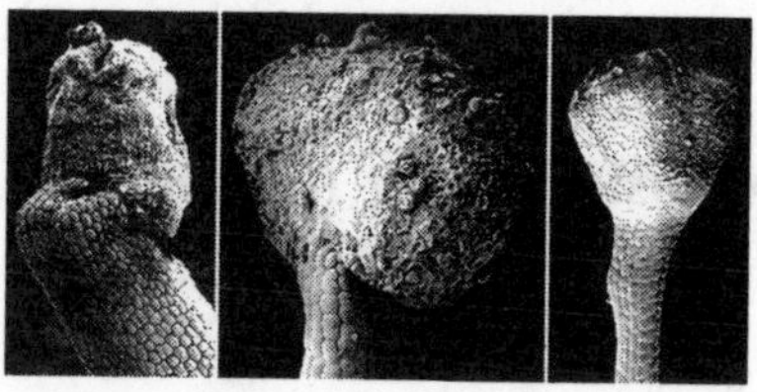

attached by its base (as certain anthers to their filaments or stalks).

- **basin**
a depression of the earth's surface into which sediments are deposited, usually characterised by sediment accumulation over a long interval, a broad area of the earth beneath which layers of rock are inclined, usually from the sides toward the centre.

- **basin cultivation**
a type of cultivation system developed to minimise erosion resulting from torrential rainfall. It involves dividing the field into many small areas each surrounded by low earth ridges that reduce run-off and allow water to infiltrate into the soil.

- **basin irrigation**

a method of irrigation wherein water is flooded into basins surrounded by earth banks.

- **basin listing (tie ridging)**
the ploughing or furrowing of land, usually on the contour, with a special lister that places earth plugs at intervals in the open furrows.

- **basipetal**
of or relating to the development or maturation of tissues or organs or the movement of substances, such as hormones, from the apex downward the base.

- **basket planting**
planting young trees in the loosely woven baskets, in which they have been sown or transplanted.

- **bat guano**
manure made from bat faeces that has been partially composed and is made of variable composition.

- **batch**
a quantity of a harvested crop put into a bin or container on a repetitive basis specifically for treatment, such as drying. The container may be utilised for storage of material after drying in contrast to continuous operation.

■ **batt**
an edible material that is attractive to the pest, which normally contains a pesticide unless used as a prebait.

■ **baule unit**
baule suggested that the unit of fertiliser, or any other growth factor, be taken as that amount necessary to produce a yield that is 50 per cent of the maximum possible. According to this concept, one baule of a growth factor is equivalent to one baule of any other growth factor in terms of growth promoting ability.

■ **bc soil**
a truncated soil profile the A horizon of which has been removed.

■ **bca horizon**
a horizon enriched in calcium.

■ **beach**
the part of a coast that is washed by waves or tides, which cover it with sediments of various sizes and composition, such as sand or pebbles. Coastal beaches occur along the coasts of oceans and seas, Lake beaches occur along the shores of lakes or large ponds. Old beaches are no longer being washed and reworked by waves but retain their original form. Subclasses include Coastal beaches (sandy) and Coastal beaches (cobbly). Beaches support little or no vegetation and have no agricultural value, although they may be sources of sand and gravel.

■ **bed**
1. a plot of ground in which plants are growing.
2. the bottom of a channel.

■ **bed method**
method of levelling the land and preparing small beds surrounded by bunds. Water is allowed to flow through the channels formed between two adjacent rows of beds.

■ **bed rock**
the solid rock beneath the soil and superficial rock. A general term for solid rock that lies beneath soil, loose sediments, or other unconsolidated material.

■ **bed up**
to build up beds or ridges with a tillage implement.

■ **bedding land**
ploughing, grading or otherwise elevating the surface of fields into a series of parallel beds or 'lands' with shallow

surface drains separating them.

- **bedding of soils**
  when the surface is drained through the provision of shallow drains formed by cultivating the land into beds (by ploughing in narrow strips or lands) — the dead furrow is left open to serve as a water course.
- **bedrock**
  the solid rock that underlies any unconsolidated sediment or soil. Limestone and sandstone are common types of bedrock in Kansas. Most Kansas bedrock is in formations of Cretaceous age or older.
- **beer's law**
  the law which states that the absorption of light by a solution changes exponentially with the concentration, all else remaining the same.
- **beet sugar (sucrose)**
  obtained from the sugar beet: chemically identical with cane sugar.
- **behaviour**
  the way of responding to a given situation.
- **beijerinck**
  the term used for aerobic nitrogen fixing bacteria. These organisms are tolerant to acidic conditions. Their presence in soil is dependent on the presence of certain plant residues rich in soluble sugars.
- **belanger's critical flow**
  that flow in open channels at which the energy content of the fluid is at a minimum. The velocity head is one-half the depth of flow and one-third the specific head if the vertical pressure distribution is hydrostatic and the velocity is uniform throughout the flow.
- **belanger's critical velocity**
  the velocity in a channel at which the energy content of the fluid is a minimum the velocity of a gravity wave.
- **belt**
  1. an elongated region where a specific condition is found.
  2. a broad strip of land characterised by one dominant feature, such as a crop or climatic type.
- **belt of no erosion**
  the upper part of hill-slope, extending from the crest for some distance down-slope, on which the surface accumulation of water from precipitation is insufficient to cause erosion.

- **belt transect**
refers to a piece of land that is taken for experimental studies.

- **bench**
a land form that is narrow and is characterised by a flat surface with a steep slope at the back. Benches may be erosional in origin, the most common being produced by lakes or rivers that have experienced a change in base level.

- **bench terrace**

a land characterised by a steep drop on the downhill side in areas with steep slopes. The main purpose of the bench terrace is to enable farming in a hilly area.

- **bench terrace water harvesting system**
the system uses most of the principles of runoff farming to conserve and utilise runoff for grain production. The conservation bench terrace system uses level contour benches with terrace ridges to control erosion and retain, spread and infiltrate storm runoff from cultivated contributing areas that are not treated to increase runoff.

- **bench-mark**
the term refers to a naturally occurring or an artificial object, bearing a marked point whose elevation above or below an adopted datum (such as sea level) is known. The usual designation is BM or PBM (Permanent Bench-Mark). A temporary or supplemental bench-mark (TBM) is of a less permanent nature and the elevation may be less precise.

- **benedict's solution**
deep-blue alkaline solution used to test for the presence of the aldehyde functional group, -CHO. The substance to be tested is heated with Benedict's solution, formation of a brick-red precipitate indicates presence of the aldehyde group.

- **beneficial water use**
beneficial water use shall be held to mean the use of such quantities of water when reasonable intelligence and reasonable diligence are exercised in its application for a lawful purpose, as is economically

necessary for that purpose.

- **benzene ring**
  it is a closed chain of 6 carbon atoms with hydrogen atoms attached with six-carbon ring structure found in $C_6H_6$, and in organic compounds formed from benzene by replacement of one or more hydrogen atoms by other chemical atoms or radicals.

- **best tillage range (soil moisture)**
  tilling a too wet soil creates puddling of the soil. When it is too dry, it has such a high degree of cohesion that it breaks into large clods and powder so that its structure suffers. At tilling, the soil should have enough consistencies to hold the tillage implement and break along the natural cleavage planes. The pH range 2.8—4.8 is found optimum for tillage operation in medium textured soils. This range is however narrow for heavy and wider for sandy soils. A soil containing good amount of organic matter could be tilled at moisture contents higher than pH 2.8.

- **beta decay**
  the process of radioactive decay in which a neutron loses a beta particle, which is physically identical to an electron. This increases the atomic number of the atom one by one turning the neutron into a proton. The atom's atomic mass number stays the same because the total number of protons and neutrons remain the same.

- **beta emitter**
  when the radioactive nuclide decays as a beta decay with the emission of a beta particle.

- **beta error**
  the acceptance of a null of other hypothesis when it is really false. Also known as error of the second kind.

- **beta factor**
  the ratio of the air saturation value in an activated sludge in pure water at the same temperature and pressure.

- **beta particle**
  an electron of either positive charge (ß+) or negative charge (ß-), which has been emitted by an atomic nucleus or neutron in the process of a transformation. Beta particles are more penetrating than alpha particles but less than gamma rays or x-rays.

- **beta ray**
  radiation of beta particles during radioactive decay.

- **beta ray spectrometer**
an instrument used to measure the energy distribution of beta particles.

- **beta regression coefficient**
the weight assigned to or the coefficient of one of the independent variables in a partial regression equation, when the variables are expressed in terms of standard scores, indicating the net contribution of that variable to the dependent variable.

- **betatron**
a particle accelerator in which magnetic induction is used to accelerate electrons.

- **between group variance**
in analysis of variance the difference between (or among) the means of the several groups involved, commonly the numerator of the F ratio in analysis of variance.

- **between habitat specialisation**
restriction of the distribution of a population to a narrow range of habitats.

- **between race**
a race of plants which is intermediate in character between the typical species and one of its well marked subspecies or varieties.

- **bh horizon**
a type of soil that is rich in humus.

- **bhusa**
the left afters of the threshing grain crops, in particular of wheat and pulses, chaffed straw of cereals.

- **bias**
the systematic tendency of any factors associated with the design, conduct, analysis and evaluation of the results of a clinical trial to make the estimate of a treatment effect deviate from its true value.

- **bibulous**
a substance that is capable of absorbing water.

- **biennial**
1. longing for two consecutive years.
2. A plant that requires two years to complete its life cycle from seed germination to seed production and death.

- **bilateral monopoly**
a bilateral monopoly is a market situation in which there is both a monopoly on the selling side and a monopoly on the buying side.

- **bilayer**
a double layer of lipid molecules such as that formed on

the surface of the aqueous phase by phisoholipids in aqueous, solution.

- **bimodal**
  a distribution in which the frequency curve has two peaks. A single peak is called a mode.

- **bin**
  physical storage container such as a small corrugated or plastic parts.

- **binary sealer**
  a counting device utilising the binary system of numbers.

- **binder tobacco**
  tobacco wrapped around the filler leaves and segments of a cigar.

- **binding energy**
  the energy with which a particle is held to an atom or nucleus. Electron binding energy is a synonym for ionisation potential. Nuclear binding energy for a neutron, proton or alpha particle is equal to the difference in mass between the original nucleus and the sum of the product particles.

- **bio**
  prefix meaning 'life'.

- **bioaccumulation**
  buildup of some specific organic or inorganic compound within a specific tissues of a given organism usually applied to certain heavy metals, pesticides or metabolites.

- **biochemical oxygen demand (BOD)**
  bioaccumulation denotes the accumulation of a substance in a living organism as a result of its intake both in the food and also from the environment. Determination of the β-factor (Bioaccumulation Factor) is extremely important in the risk analysis of a compound.

- **biochrome**
  any natural colouring matter of plants and animals, a biological pigment.

- **biocide**
  a chemical which has a wide range of toxic properties, usually to members of both the plant and animal kingdoms.

- **bioclimatic law**
  the maturity of a given crop may vary by 4 days for each 1° latitude, 5° longitude and 400″ altitude.

- **bioclimatology**
  the study of effects of climate on living organisms.

- **bioclimatology**
  the study of climate in relation to organic life, including human beings.

- **biocoenosis**

a community or natural assemblage of organisms. The term often is used as an alternative to ecosystem, but strictly it is the fauna/flora association excluding physical aspects of the environment.

- **biocoenosium**

where both plants and animals can live.

- **bioconcentration (bioamplification)**

accumulation of a substance in an organism by absorption from the environment irrespective of any intake with food. The concept is of particular importance for aquatic life with regard to the absorption of those fat-soluble substances which are only broken down slowly.

- **biodemography**

the study of the integration of ecology and genetics of populations.

- **biodynamics**

the study of the active vital phenomena of organisms.

- **bioenergetics**

a method of studying and understanding the human personality in terms of the body and its energetic processes.

- **biofertilisers**

are preparations containing live or latent cells of efficient strains of nitrogen fixing, phosphate solubilising or cellulotyic microoganisms and accelerate certain microbial processes to augment the extent of the availability of nutrients in a form which can be easily assimilated by plants.

- **biofloccuzlation**

1. the aeration of settled [sewage with activated sludge for 1 h, followed by treatment of the effluent on trickling filters. The activated sludge is re-aerated.
2. flocculation of a suspension containing bacteria, algae, etc.
3. flocculation of activated sludge in a secondary sedimentation tank.

- **biogcochemical cycling**

the cycling of chemical constituents through a biologic system.

- **biogen**

an imaginary protein molecule which is unbalanced, but has the essential properties of a living thing.

■ **biogenetic law**
the hypothesis that, during its development an individual repeats the evolution of its race in a shortened form.

■ **biogeochemical cycle**
the cycling of such chemical elements as nitrogen, carbon, and sulphur from the environment into organic substances and back into the environment.

■ **biogeography**
the scientific study of the geographic distribution of organisms.

■ **biologic oxidation**
the process by which bacterial and other micro organisms feed on complex organic materials and decompose them. The self-purification of waterways and activated sludge and trickling filter waste water treatment processes depend on this principle. The process is also called biochemical oxidation.

■ **biological clock**
the internal mechanism of an organism that regulates circadian rhythms and various other periodic cycles.

■ **biological control**
control of pest populations by natural processes or naturally-occurring organisms.

■ **biological half life**
the use of one organism to control the population size of another organism.

■ **biological indicator (or index) of pollutant**
the presence and frequency of a plant, animal, microbe or other form of life can give an idea of the level of pollution in its environment. An organism that indicates pollution may provide no indication about poisons. Various classifications of organisms indicating levels of organic pollution have been produced for e.g., the saprobic classification, the Trent biotic index.

■ **biological insecticide**
a biological agent such as *Bacillus thtiringiensis*, which kills insects like a chemical insecticide, and then rapidly dissipates the environment.

■ **biological interchange**
the interchange of elements between organic and inorganic states in a soil of other substrate through the agency of biological activity. It results from biological decomposition of organic compounds and the liberation of inorganic ma-

terials on the one hand (mineralisation) and the utilisation of inorganic materials in the synthesis of microbial tissue on the other (immobilisation). Both processes commonly proceed continuously in normal soils.

- **biological mineralisation**
  the conversion of an element occurring in organic compounds to the inorganic form as a result of biological decomposition.
- **biological monitoring**
  the use of a biological entity as a detector and its response as a measure to determine environmental conditions. Toxicity tests and biological surveys are common biomonitoring methods.
- **biological quality**
  the sum of the individual factors present in the plant that contribute towards maintaining normal metabolism in the organism of the living being, animal or human, consuming that plant.
- **biological spectrum**
  a table showing the percentage frequency of occurrence of the various plants and animals in a defined area.
- **biological weathering**
  refers to physical and chemical weathering assisted by biological agencies.
- **biological yield**
  total dry matter produced by a plant or of a crop in unit land area.
- **bioluminescence**
  production of light by a chemical reaction within an organism. Bioluminescence generated by firefly luciferase is a sensitive assay for ATP, which is utilised during the reaction.
- **biolysis**
  the decomposition of organic matter resulting from activity of living organisms, disintegration of life.
- **biolytic**
  a substance or material that can destroy life.
- **biomass**
  an energy resource derived from organic matter. These include wood, agricultural waste and other living-cell material that can be burned to produce heat energy. They also include algae, sewage and other organic substances that may be used to make energy through chemical processes.

- **biomass density**
weight of living material expressed as a dry weight, in all or part of an organism, population, or community expressed as weight per unit area.

- **biome**
the complex of living communities maintained by the climate of a region and characterised by a distinctive type of vegetation.

- **biometer**
a method of assessing the amount of life by measuring the respiration for e.g., in a soil sample.

- **biometry**
the study of the statistics of biological phenomena, the application of mathematics and statistics to the study of living things.

- **biomonitoring**
1. the use of living organisms to test the suitability of effluents for discharge into receiving waters and to test the quality of such waters downstream from the discharge.
2. the biological treatment of wastewater and sludge, by inducing the breakdown of organics and hydrocarbons to carbon dioxide and water.

- **bion**
1. a hypothetical unit of life
2. a general term for a living cell or organism.

- **bionomics**
the branch of biology concerned with the relations between organisms and their environment.

- **biont**
a discrete unit of living matter.

- **biopedology**
the study of the living beings (plant and animal) of the soil and their influence on it.

- **biophysical**
the conditions pertaining to the water conditions of the environment of the living organisms.

- **biopoiesis**
the origin of organisms from replicating molecules. Biopoiesis is a corner stone of a biogenesis. Deoxyribonucleic acid (DNA) is the best example of a self-replicating molecule and is found in the chromosomes of all higher organisms. In some bacteria ribonucleic acid (RNA) is self-replicating. Various chemical and physical conditions must be met before, either DNA or RNA is able to replace.

- **bios factor**

  refers to essential substances that contribute to the growth of a plant, especially yeasts obtained from the environment. It is a mixture of aneurin, biotin and other substances.

- **biosequence**

  a sequence of related soils that differ, one from the other, primarily because of differences in kinds and numbers of soil organisms as a soil forming factor.

- **bioseries**

  the change in the historical sequence in any one inheritable character formed during the evolution.

- **biosphere**

  the environment of living organisms penetrating and interacting with the lithosphere (rock minerals and soil), hydrosphere (water), and atmosphere (air). Included in living organisms are plants and animals from low-growing grasses to giant trees and from microscopic microbes to greedy goats.

- **biostabiliser**

  a device that grinds and aerates the solid wastes into compost.

- **biosystematics**

  the area of systematics in which experimental taxonomic techniques are applied to investigate the relationships between taxa. Such techniques include serological methods, crossing experiments and cytological examination, in addition to the more established procedures of comparative anatomy. Evidence from ecological studies may also be brought to bear.

- **biotape**

  1. a region where the original conditions and the adapted conditions are uniform.

  2. an area where the principal habitat conditions and the living forms adapted to the conditions are uniform.

- **biotic**

  refers to the effect of living organisms in the development of vegetation and soils.

- **biotic adaptation**

  the change in the form or physiology, because of the competition with other plants.

- **biotic climax**

  a climax community maintained by living organisms for example, grassland remaining static due to grazing of animals.

- **biotic complex**
the adaptation of the soils, plants and animals according to the climatic and other environmental conditions.

- **biotic control**
the natural control of the numbers of an insect, animal or plant by its enemies and diseases.

- **biotic factor**
the factors influenced by any living organisms. Usually restricted to the influence of animals including man.

- **biotic potential**
the inherent power of an organism to multiply over a given period of time in the absence of control factors.

- **bird repellent**
1. when birds are discouraged or driven away with the use of poisonous substance.
2. a substance which drives away birds or discourages them from roosting.

- **bird scarer**
mechanical device used for scaring away birds, produces loud reports at regular intervals, operates on calcium carbide, and is perched on treetops, can scare birds in an area of 4 acres. One person can control 4 machines at a time, as they only need replacement of carbide and water.

- **birth of a soil**
the process of the birth of a soil starts the moment a portion of the parent rock is exposed to the action of the atmosphere and living organisms.

- **bisect**
a profile of plants and soil showing the vertical and lateral distribution of roots and tops in their natural position.

- **bitumen**
refers to the substance that are extracted from the soil and are insoluble in water and soluble in the organic solvents. Soil bitumens are mixtures of fatty combustible substances or hydrocarbons: waxes, gums, resins, fatty acids, poor in oxygen and rich in carbon.

- **bituminisation**
when the organic substances are transformed into bituminous matter.

- **bituminous material lined channel**
refers to the irrigation channel lined with a mud plaster containing a small quantity of bituminous material to sides and bottom. Gives a high degree of waterproofing with

little outlay of money.

- **biuret reaction**

1. a chemical reaction in which an alkaline solution of biuret gives a purple colour on the addition of cupric sulphate.
2. used as a biochemical test for protein and urea.

- **biuret test**

a standard test for proteins and their derivatives, for e.g., peptides and peptones. Sodium hydroxide is first mixed with the test solution and then copper (II) sulphate solution is added drop by drop. A violet colour indicates a positive result. The reaction is due to the presence of — NH— CO—in the molecule, the colour not appearing with free amino acids. The compound biuret also readily forms from urea around 150°—170°C and thus the biuret test can be used to identify urea.

- **black alkali**

also alkaline soil, when a brown or black incrustation is left after sodium carbonate is dissolved into an organic matter on the surface of evaporation of moisture.

- **black body**

1. a substance which absorbs all the radiation falling on it. Its absorptance and emissivity are both equal to 1.
2. a perfect black body is one which absorbs totally all the radiation of any wave length which fall on it.

- **black cotton soil**

soil of black colour (throughout the profile) with low organic matter and a high clay content, exhibits a strong phenomenon of swelling and shrinkage, during dry periods cracks develop that may be several cm wide and more than 50 cm deep. Depending upon their depth the black cotton soils are classified as shallow, medium and deep black soils (chromusterts).

- **black earth**

a terms given to the soil that has a high quantity of chernozem and dark clay of tropics .

- **black market**

transactions in goods and services that violate controls on price or supply imposed by the government or producer. Such transactions are usually illegal and occur at a price much higher than the competitive equilibrium price.

- **blade harrow (bakhar)**

a frame with spikes or sharp-edged disks, drawn by a horse

or tractor and used for breaking up and leveling plowed ground, covering seeds, rooting up weeds, which consists of one or more blades attached to the beam or frame used for shallow working of the soil with minimum of soil inversion. It is used to prepare seed beds mostly in clay soils. It works like a sweep which moves into the top surface of the soil without inverting the soil.

- **blanket application**
an application of spray or granules over an entire area rather than only in rows, beds, or middles.

- **blast furnace slag**
a byproduct of the manufacture of pig iron. In the reduction of iron, the $CaCO_3$ in the charge loses its $CO_2$ and the CaO then combines with the molten silica to form a slag that is tapped off and either air cooled or quenched with water. It is then powdered, screened and bagged. It is used as a liming material for acid soils.

- **blastogenesis**
transmission of inherited characters by the germplasm only.

- **bleaching**
1. loss of colour, usually caused by exposure to sunlight.
2. any process which lightens the colour in a plant by the destruction of the chlorophyll or from preventing the green colour from developing by covering from sunlight.

- **bleicherde**
the light-coloured, leached, A2 horizon of Podzol soils.

- **blending inheritance**
the inherited characteristics of the offsprings that bridges the original parents and the successive generations. This is because several genes control the character.

- **blending of water supply**
it is a method of improving the quality of saline irrigation water. If an alternative supply is available, but not adequate in quantity, a blend of waters may offer an overall improvement in quantity and quality of irrigation water. Blending of water is a simple but drastic solution to a high EC and sodium toxicity problems. Since the proportions of monovalent and divalent cations adsorbed on the soil are concentration dependent, and that the dilution favours adsorption of cations of highest valence such as calcium and

magnesium over sodium. Any change in quality due to blending may be evaluated from the following relationship. Resulting EC of mix=EC of source Ax proportion of A+EC of B X its proportion.

- **blight**
common name for a number of different diseases on plants characterised by the rapid death of plant tissue e.g. leaf blight, blossom blight, shoot blight.

- **blind cultivation**
cultivating with a harrow, weeder, rotary hoe, or other implement to kill weeds before a seeded or planted crop has come up.

- **bloat**
an excessive accumulation of gases in the rumen of animals.

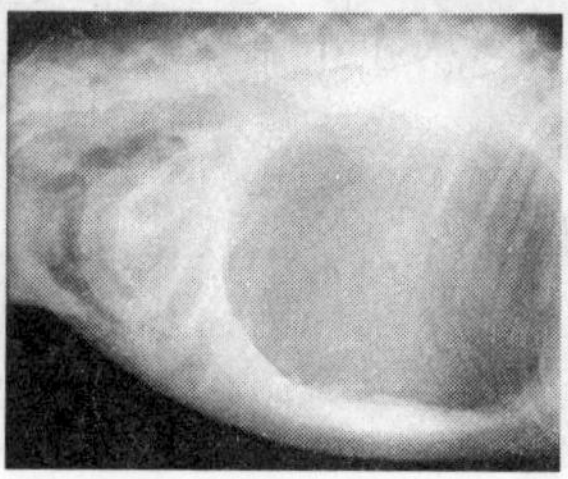

- **block**
a sector or group of sectors that function as the smallest data unit permitted.

- **blown-out land**
1. consists of areas from which all or most of the soil material has been removed by wind—a condition resulting from an extreme degree of soil blowing or wind erosion. The areas are shallow that has flat or irregular floors formed by some more resistant layers, by an accumulation of pebbles or cobbles, or by exposure of the water table. Some areas have a small proportion of hummocks or small dunes. The land is barren, or nearly so, and generally useless for crops.
2. Transient areas of Blown-out land, developed in loose deep sand, are included along with adjacent dunes in Sand-dune land. Small areas of Blown-out land are often called 'blowouts' and are shown with symbols.

- **blue green algae**
prokaryotic organisms with a bacteria-like cell structure, lacking a nucleus and other organelles, these species manufacture photosynthetic pigments but lack chloroplasts, the specialised photo

synthetic organelles in higher plants, in some situations an increase in blue-green algae can indicate an environmental stress such as pollution.

- **bluff vegetation**

includes the lithophytes fixed directly on the surface of the rock, and the chasmophytes which grow into the cracks. Contribute to the disintegration of the rock and to the formation of the lithosol.

- **body burden**

the amount of a radioactive material present in a body over a long time period. It is calculated by considering the amount of material initially present and the reduction in that amount due to elimination and radioactive decay.

- **bog iron ore**

impure ferruginous deposits developed in bogs or swamps by the oxidising action of algae, bacteria, or the atmosphere on iron carried in solution.

- **bog soils**

an intrazonal group of soils with a muck or peaty surface underlain by peat, developed under swamp or marsh types of vegetation, mostly in humid or subhumid climate.

- **bohr radius**

the orbital radius (cm) calculated from the Bohr equation: $a = n^2h^2 / 4n^2mZe^2 = (0.529 \times 10^{-8})n^2$ where n is the principal quantum number.

- **bohr's effect**

an effect by which an increase of carbon dioxide in the blood and a decrease in pH results in a reduction of the affinity of hemoglobin for oxygen.

- **boiling point**

the temperature at which the equilibrium vapour pressure between a liquid and its vapour is equal to the external pressure on the liquid. Compare ice point.

- **boll**

the fruit of the cotton plant.

- **bone ash**

a white ash consisting primarily of tribastic calcium phosphate obtained by burning bones in air.

- **bone char**
black substance containing char in the form of carbonised bone also used as a black pigment and contains 30 to 35% $P_2O_5$.

- **bone meal**
it is obtained by processing the bones of dead animals whereby glue and fats are separated from the crushed material. The crushed material is bagged and utilised as by-product meal fertiliser material. Bone meal is also used as cattle feed.

- **bone phosphate of lime (BPL)**
is the phosphate content of phosphate ore expressed as tricalcium phosphate $Ca_3(PO_3)$.

- **bone tankage**
is a concentrated organic manure containing 3 to 10% N and 7 to 20% $P_2O_5$.

- **bonitation**
the evaluation of the numerical distribution of a species in a particular locality or season, in relation to agricultural, veterinary or medical implications.

- **booking**
it is the method of recording the ground measurement in the field book.

- **boom**
a situation in which production is above long-run trend and has been growing rapidly, in which employment is high and unemployment is low, and in which nearly everyone is optimistic about the future of the economy.

- **booster pump**
a surface pump used to increase pressure in a water line, or to pull from a storage tank and pressurise a water system.

- **boot**
1. the upper leaf sheath of a grass.
2. the stage at which the inflorescence expends the boot.

- **boot stage**
the growth stage of grasses at the time the head is enclosed by the sheath of the uppermost leaf.

- **borax ($na2b_4o_7.10h_2o$)**
contains about 11% boron. Used in fertiliser as a source of the boron, a plant food element.

- **bordeaux mixture**
a fungicide made of copper sulphate and quick lime. A common formula is 4 Ib copper sulphate, 4 Ib quick lime, dissolved in 50 gallons of water.

- **bordeaux paste**
fungicide made by taking one pound of copper sulphate and dissolving it in one gallon of water, similarly slaking one pound of quicklime and making it up to one gallon by adding water, and then pouring the copper sulphate solution into the lime solution. Resulting mixture has consistency of paint, and is applied to the bark and cut ends of plants with brush.

- **border**
1. a strip of land, varying in width and length edged on both sides by a low ridge, which guides the flow of water within that strip.
2. trees or shrubs planted along the border or boundary.

- **border irrigation (strip irrigation)**
the application of water to strips marked by levees 10 to 100 feet apart. The soil is essentially level between levees, on even topography, with a gentle slope so the water forms a sheet over the whole area.

- **border parenchyma**
a sheath of one or more layers of parenchymatous cells surrounding a vascular bundle.

- **border rows**
in hybrid seed plot, recommended number of rows of the male parental line grown on all the sides of the field.

- **border strip**
a grassed thickly vegetated strip at the edge of a field, along outlet channels, or at ends of rows to check or prevent erosion.

- **border strip cropping**
a kind of strip cropping involving growing of strips of permanent vegetation placed around the border of a field, a farm, or a body of water. The border strip may be used to control either wind or water erosion. It is espe-

cially useful to filter the silt out of runoff water before it reaches a pond or stream. Border strips are very useful when placed between feedlots and streams.

- **boron**

one of eight micronutrients. Boron is believed to be involved in carbohydrate transport in plants, it also assists in metabolic regulation. Boron deficiency will often result in bud dieback. Boron is associated with the reproductive phase in plants and its deficiency is often found to be associated with sterility and malformation of reproductive organs. It is also considered to be involved in carbohydrate metabolism, particularly in the translocation of photosynthates—the sugars. Boron is absorbed in one or more of its ionic forms, such as $H_2BO_3$, $HBO_3^2$.

- **boron counter**

an ionisation chamber containing a boron compound for the detection of neutrons.

- **boron deficiency indicator plant**

root crops like table beet, turnip and mangel are good indicators of boron deficiency. Out of the leafy vegetables, cabbage, cauliflower, Brussels sprouts, and celery are the best indicator plants. In the group of forage crops, alfalfa, clover and white clover show characteristic leaf symptoms. Apple and pear are the most sensitive in the fruit group.

- **boron deficiency symptoms**

the symptoms vary from species to species but, in general, boron deficiency leads to the degeneration of the meristematic tissue (including the cambium) and to the breakdown of the walls of parenchyma cells and further to feeble development of the vascular tissue, phloem and xylem.

- **boron silicate glasses**

materials formed by fusing borax with silicate glass and shattered. They are called frit and are used as fertiliser material. Their boron content varies from 3 to 6 per cent and boron availability varies with the particle size. The finer the particle, the greater the availability because of the increase in surface area. The advantage of these compounds lies in their low solubility and the slow availability of boron, which reduces leaching losses and helps particularly the long duration crops, like apple and alfalfa, which are sensitive to boron deficiency.

- **boron toxicity symptoms**
young leaves develop chlorosis between the veins, which starts from the base of a leaf and spread towards the apex. The leaf margins and the chlorotic areas develop diffuse-brown or reddish-brown areas. Such areas may also be found on the leaf-sheaths, petioles and stem, but are often more numerous and conspicuous in the areas between the veins alongside the veins which often turn bluish-green. In some cereals, for e.g., barley and maize, the emerging leaves may be severely chlorotic or bleached and may fail to unroll. Blossoms are appreciably depressed and flower buds may show chlorotic sepals and petals.

- **boron water**
water with or without the conditions of salinity and alkalinity but containing limiting or excess of boron as well.

- **bottom contraction**
the vertical distance from the weir crest to the bottom of the weir pond.

- **boulder clay**

poorly sorted, unstratified sediment, with grains ranging in size from rock 'flour' to boulders, deposited beneath glaciers and ice sheets, or from melting ice.

- **bound water**
water chemically tied to food in the form of hydrates of inorganic salts or inorganic substances. The hydrophilic surfaces of proteins and polysaccharides have bound water on them, increasing their apparent surface area.

- **bounty**
a subsidy paid by the government to the exporters of certain products to enable them to be sold more cheaply abroad, thereby stimulating the exports of such products.

- **bouquet stage**
when the chromosomes lie in

loops with their ends near one part of the nuclear membrane during cell division.

- **bouyoucos hydrometer**
a types of hydrometer that is used in mechanical analysis.

- **bouyoucos method**
a method of measurement of soil moisture using the variations of electrical resistance of a gypsum block placed in contact with the soil.

- **brace root**
an aerial root that functions to brace the plant, as in corn.

- **brackish**
having a salinity between that of fresh and sea water.

- **brackish water**
water that is neither falls in the category of salt water, nor in the category of fresh water. It holds the middle between either one of the categories.

- **bracteole**
A small, sometimes scale-like leaf, usually associated with flower clusters.

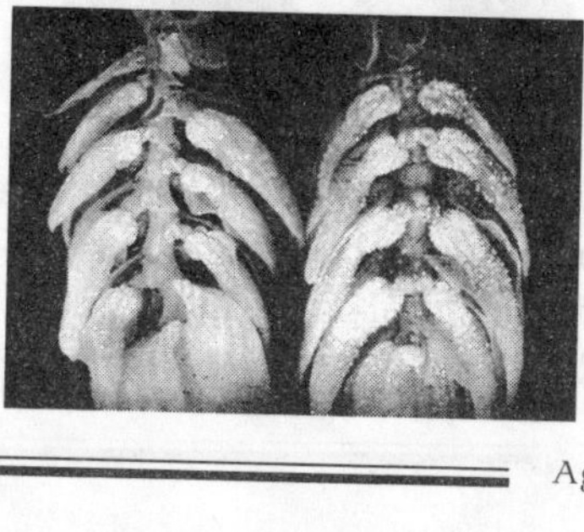

- **brady**
originally form a Greek prefix, 'bradys' means 'slow'.

- **brak soil**
an alkaline soil found in South Africa. The dominating method of formation is the extreme temperature changes.

- **brake horse power**
horsepower required at the blower, compressor, fan, or pump shaft to perform actual work. Often less than the electric motor nameplate rating, but not necessarily.

- **branching**
the occurrence of two or more modes by which a radionuclide can undergo radioactive decay. For example, RaC can undergo a and b decay. 64Cu can undergo b-, b+, and electron capture decay. An individual atom of a nuclide exhibiting branching disintegrates by one mode only. The fraction disintegrating by a particular mode is the branching fraction for that mode. The branching ratio is the ratio of two specified branching fractions.

- **brand**
1. a name, term, design, symbol, or any other feature that identifies one seller's good or

service as distinct from those of other sellers.

2. The legal term for brand is trademark. A brand may identify one item, a family of items, or all items of that seller.

- **brashy soil**

soil composed almost entirely of coarse, angular rock fragments, with fine soil matter filling the interstices.

- **braunification**

release of iron from primary minerals and the dispersion of particles of iron oxide in increasing amounts, their progressive oxidation or hydration, giving the soil mass a brownish colour.

- **breaking strength**

a loose term applied to a given structural member with reference to the ultimate load it can sustain under a set of given conditions.

- **breaking stress**

the ultimate stress required to cause a complete failure in a test specimen, or any structural member under the action of bending.

- **break-through**

that point in an adsorption cycle at which the effluent begins to show a marked increase in concentration of the substance being adsorbed.

- **breathing root**

a root produced by large plants growing in mud, it projects above the mud and water allowing the air to reach the root below.

- **breccia**

derived from the Latin meaning 'broken', is a rock formed of angular fragments in a matrix which may be of a similar or a different material.

- **breeder reactor**

a nuclear reactor that produces fissionable material faster than it consumes it. The new fissionable material is created by neutron capture in fertile material by a process known as breeding.

- **breeder's seed**

1. seed stock produced by the breeder or by the agency licensed to produce or maintain 2. a variety,

source of foundation seed.

- **bridging host**
  an intermediate host, whereby a specialised parasite passes from a susceptible to a resistant host.

- **brine**
  is water containing relatively large amounts of salts. Brine water is created in areas of strong evaporation, like the Dead Sea, and in polar regions, when salt is ejected from water during freezing.

- **brix**
  the percentage of total solids in sugarcane juice, read off from brixometer, ranges from 18 to 22, 15 in juice of young and immature cane.

- **brix saccharometer**
  an instrument used for measuring the degree of ripeness and gives the direct reading of the total solids dissolved in the juice. It varies from 20 to 22.5 depending on the variety.

- **brixometer**
  a special hydrometer calibrated to read off directly the quantity of total solids in sugarcane juice.

- **broad base**
  a system of crop growing in wide bands with wide dikes. The water is channelled to the bottom of the slope by a canal.

- **broad base ridge terrace**
  type of contour terrace for soil conservation, wide, low embankment formed on contour by excavating soil from both sides of it, more suited to low rainfall areas and fairly absorptive soils.

- **broad base terrace**
  a low, wide terrace used on gentle slopes.

- **broad leaf species**
  in botany, those plants classified as dicotyledoneae, morphologically—those having broad, rounded or flattened leaves as opposed to the narrow blade-like leaves of the grasses, sedges, rushes and onions.

- **broad leaved tree**
  1. trees belonging to the botanical group called dicotyledons, and producing timber usually known as 'hardwood' floodway.
  2. a channel built to divert the ordinary flow from a particular reach of the main channel for any reason.

- **by wash channel**
  a channel to reduce the amount of silt coming into a reservoir from direct overland

flow. It is a least or stream following a contour, which starts at the head of the reservoir, passes along the full length of the lake and discharges to the spillway. Least may be cut along each of the two hillsides flanking the lake.

- **by-pass gate**
  a relatively small gate through which water is passed to equalise pressure on both sides of a large gate.

- **by-pass valve**
  a small pilot valve used in connection with a large valve to equalise the pressure on both sides of the disk of the larger valve before the larger valve is opened.

- **by-product**
  a substance obtained incidentally during the manufacture or production of some other substance. For example, nitrites, after oxygen has been removed from nitrates. Oxygen is a by-product of photosynthesis.

- **c horizon**
  soil horizon normally found below the B horizon and above the R horizon. This layer is composed of weathered bedrock that has not been yet significantly affected by the pedogenic processes.

- **$c_3$ plant**
  plants whereby the first organic product of carbon-fixation (during the Calivin cycle) is a three-carbon compound. This category includes the vast majority of common plant species.

- **$c_4$ plant**
  plants whereby the first oragnic product of carbon-fixation is a four-carbon compund, which is formed prior to the Calvin cycle. This adaptation is advantageous in hot regions with intense sunlight as it minimises photorespiration and maximises sugar production. Sugarcane is an example of this type of plant. $C_4$ plants are more efficient than $C_3$ plants: they are capable of utilising much higher light intensities and temperatures, have up to double the maximum rate of photosynthesis, and lose less water by transpiration because with more efficient $CO_2$ uptake, smaller stomatal apertures are needed to obtain sufficient $CO_2$ for photosynthesis.

- **CA**
  a white metallic element that burns with a brilliant light, the fifth most abundant element in the earth's crust, an important component of most plants and animals

**caeruleus**
a Latin word that means 'pale blue'.

**caffeine (theine $c_8hi_0o2n_4$)**
a naturally-occurring substance found in the leaves, seeds or fruits of over 63 plant species worldwide and is part of a group of compounds known as methylxanthines. The most commonly known sources of caffeine are coffee and cocoa beans, cola nuts and tea leaves. Caffeine is a pharmacologically active substance and depending on the dose, can be a mild central nervous system stimulant. Caffeine does not accumulate in the body over the course of time and is normally excreted within several hours of consumption.

**calcareous**
covered with or containing lime.

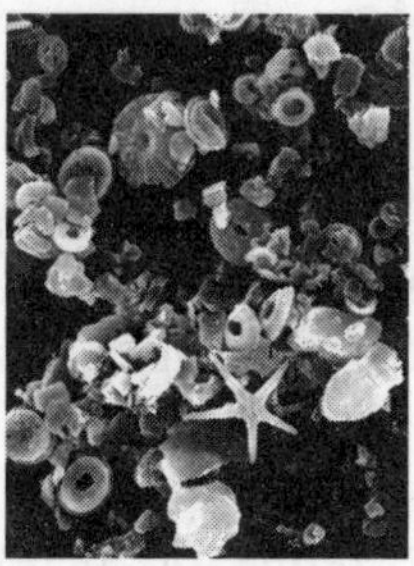

**calcareous pan**
a hard layer of limey material, impermeable to water, formed below the surface of the soil, and affecting the water-supply of plants growing above it.

**calcareous soil**
1. a soil containing enough calcium carbonate (commonly with magnesium carbonate) to effervesce (fizz) visibly when treated with cold, dilute hydrochloric acid.
2. A soil having measurable amounts of calcium carbonate or magnesium carbonate.

**calcic amelioration**
calcium enrichment by addition of Hme, limestone or marl powder to the soil.

**calcic horizon**
a horizon of secondary carbonate accumulation more than 6 inches in thickness, has a $CaCo_3$ equivalence of more than 15 per cent and at least 5 per cent more $CaCO_3$ than the C horizon.

**calcicole**
a plant that thrives on a calcareous soil.

- **calcicolous**
growing or living in soil and rich in lime.

- **calcicolous plants**
are those which are found primarily in calcareous land.

- **calcification**
1. is a soil forming process which occurs in arid and semi-arid climatic zones, and leads to the deposition of calcium carbonate in the B-horizon.
2. the accumulation of calcium carbonate, on or in the cell-wall.

- **calcifuge**
1. a plant that thrives on an acid soil.
2. a plant that thrives only in soils poor in calcium carbonate.

- **calcifugous plants**
1. plants avoiding calcareous lands.
2. Plants which are not found or which develop badly on calcareous soil. Such plants are also called siliceous because the land over which they grow are rich in free or combined silica. Calcium carbonate is toxic to calcifugous plants. It sometimes prevents them from assimilating the iron necessary for chlorophyll.

- **calcimeter**
a device for measuring the carbonate content of soil. It measures the volume of CC>2 released as a result of addition of hydrochloric acid.

- **calcimetry**
the fact that each gram of pure $CaCO_3$ evolves 224 cubic cm of $CO_2$ gas under the action of $HC_1$ is used in the measurement of the carbonate content of the soil. The $CO_2$ so evolved is collected in the calcimeter.

- **calciphobe, calcifobous, calcifuge**
a plant species which does not thrive on a lime soil. It prefers an acid soil.

- **calciphytes**
plants that require or tolerate considerably amounts of calcium or, are associated with soil rich in calcium.

- **calcite**
a common rock forming mineral of calcium carbonate (CaCO3), found primarily in limestones but also commonly forming a diagenetic cement in many other rock types and as a gangue mineral in many ore deposits.

- **calcium**
element number 20 in the periodic table. It is a metal

but, in living tissue, occurs only in its reacted form, as an electrically charged ion Ca2+. The cell actively transports it across the outer membrane and this ion plays a role in determining the electrical voltage difference between the inside and outside of the cell. It also has a role in controlling muscle contraction. Plants deficient in calcium are stunted because they produce fewer and smaller cells. They have weak stems because their cell walls are less than normal thickness. The inability of calcium deficient roots to elongate rapidly handicaps the plant for exploiting new portions of the soil volume to obtain water and nutrients.

- **calcium ammonium nitrate (CAN)**

1. this is an inorganic material made by reacting ammonia with nitric acid and adding and pulverised dolomite of calcium carbonate before prilling or granulation.
2. It is one of the old solid forms of nitrogen fertiliser, ammonium nitrate, to which ground limestone has been added. It (CAN) contains around 25 % N. The pulverised limestone is added to increase stability and improve storage and handling properties.

- **calcium chloride (road salt)**

a calcium compound that is used to raise the calcium hardness of the water. Typically, available in a granular form for use in pools and a liquid form for use in spas and hot tubs.

- **calcium deficiency symptoms**

a deficiency of calcium produces striking symptoms. Meristematic regions found at stem, leaf, and root tips are greatly affected and eventually die, thus terminating growth in these organs. Roots may become short, stubby, and brown as in calcium-deficient tomato plants. Chlorosis generally occurs along the margins of younger leaves, these areas usually becoming necrotic. Malformation or distortion of the younger leaves is also characteristic of calcium-deficient plants, a hooking of the leaf tip being the most easily detected symptom. Deficiency symptoms generally appear first in the younger leaves and the growing apices, probably because of the immobility of calcium in the plant. Cell walls may become rigid or brittle.

- **calcium metaphosphate**
it is manufactured by burning phosphorus in a furnace where the hot gaseous combustion products pass through lump rock phosphate packed in the upper chamber of the furnace. The hot phosphorus vapour and rock phosphate chemically combine to form calcium meta-phosphate which contains 62-64 % $P_2O_5$.

- **calcium monosulphide (CAS)**
a fungicidal powder of varying colours. The fungicidal efficacy of the compound depends on the calcium sulphide present.

- **calculated drainage**
the water height D (mm) which infiltrates in one year in a given soil. It is related to the annual precipitation P (mm) and to the average annual temperature T (°C) by an approximate relation given below:
$D=yP^3:(l+yP^2)$
where y is the coefficient of continentality=l/ (0.15 T— 0.13).

- **calibrate**
to determine, by measurement or comparison with a standard, the correct value of each scale reading on a meter or other device, or the correct value for each setting of a control knob. The levels of the calibration standards should bracket the range of planned measurements.

- **caliche**
a more or less cemented deposit of calcium carbonate or of mixed calcium and magnesium carbonate, characteristic of soil of warm or hot desert and semiarid regions. It may be soft, thin soil horizons or hard, thick beds just beneath the solum or exposed at the surface by erosion. It results due to precipitation from the soil solution.

- **calmonite**
a naturally occurring compound of calcium borate (CazB ^ CV SH ^ O) used as fertiliser. It is considerably less soluble than borax hence superior to it on sandy soil.

- **calomel electrode**
a half cell—having a mercury electrode coated with mercury (I) chloride, in an electrolyte

consisting of potassium chloride and (saturated) mercury chloride solution. Its standard electrode potential against the hydrogen electrode is accurately known and it is a convenient secondary standard.

▪ **calomorphic soils**
are intrazonal soils occurring on permanent materials very high in lime content. They are known as rendzinas (a potash word). They occur in cool to hot and humid climate. Calcification is the main factor of development of the rendzina profiles.

▪ **calorie**
a calorie is the amount of energy required to raise the temperature of one milliliter (ml) of water at a standard initial temperature by one degree centigrade (iC). Thus 1 Cal. = 1,000 cal. , and 1,000 Cal.=l therm.

▪ **calorimeter**
a device for measuring thermal energy , for example in determining thermal capacity, specific latent thermal capacity, energies or combustion, etc.

▪ **calurea**
a dry, odourless, white-crystal sometimes referred to as calcium nitrate-urea. It contains 34 per cent nitrogen and 10 per cent calcium. The double salt has an excellent physical condition for spreading in a fertiliser distributor. If it is mixed with superphosphate, the resulting mixture has a poor physical condition.

▪ **calutron**
a device used to separate isotopes. It is based on the principle of the mass spectrometer.

▪ **calyx**

the outer whorl of a flower made up of sepals which are usually green and protect the flower in the bud.

▪ **cambic horizon**
an altered horizon resulting from movement of soil particles by frost, roots, and animals to such an extent to destroy original rock structure.

▪ **cambium**
a meristem with products of divisions arranged in orderly fashion in parallel files. Consists of one layer of initial cells and their derivatives.

- **camnium**
  a succession of plants caused by cultivation.
- **campestris**
  a Latin word that means 'growing in fields'.
- **can crushers**
  are used for crushing sugarcane and extracting cane juice for conversion into the famous rural product called gur.
- **cane carrier**
  conveys can from tip to cane crushing plant.
- **canker**
  1. a definite, relatively localised, necrotic lesion primarily of the bark and cambium.
  2. a plant disease, caused by bacteria or fungi, giving a limited necrosis of the cortical tissue.
- **canopy**
  the branches, leaves, etc. formed by woody plants, some distance above the ground. Near about a continuous stratum of foliage.

- **canopy (forestry)**
  in a forest, the cover of green leaves and branches formed by crowns of all the individual trees. Its density is ordinarily expressed as the amount (or percentage, of the ground that would be completely shaded by the forest if the sun were straight overhead).
- **canopy density**
  the relative completeness of canopy, usually expressed as a decimal coefficient, taking closed canopy as unity. The following classification of canopy density is in vogue: Closed when the density is 1.0, Dense when the density is between 0.75 and 1.0, Thin when the density is between 0.5 and 0.7, and open when the density is less than 0.5.
- **canopy volume**
  is obtained by multiplying soil area by crop height.
- **cantonment**
  the process of transferring to the holders of rights over a property, the ownership of part of the property in com-

pensation for extinguishing rights over the reminders.

- **canus**
  a Latin word that means 'grey white'.

- **capability ratings**
  the classification of land (or soil) according to its ability to produce and practices necessary to obtain good production economically.

- **capacity (pump)**
  the volume of water pumped per unit time. It is generally measured in litres per second or litres per minute or litres per hour or even cubic metres per second.

- **capacity factor**
  a measure of the available amount of the nutrient. The total quantity of an element that is transferred to the soil solution when the latter is depleted by a crop or by leaching *i.e.* the reserve that can move into the soil solution during plant growth. The chemical methods currently used for identifying the capacity factor are designed to extract the readily exchangeable or soluble nutrient, even though the true mechanisms of the supply of elements to the soil solution are not fully understood.

- **capillarity**
  is the phenomena which leads to holding of water in the soil by the forces of surface tension as tiny films around soil particles and also in the capillary spaces. This water is stretched above the height at which it could be held by hydrostatic pressure, often against the pull of gravity.

- **capillary**
  an interstice of the size of a hair. They are cylindrical tubes less than 500 microns in diameter.

- **capillary action**
  the phenomena observed in liquids due to unbalanced inter-molecular attraction at the liquid boundary.

- **capillary attraction**
  apparent attraction between soil and water, considered to cause moisture to move to drier parts of soil.

- **capillary conductivity**
  a denned as the grams of water that flow per second through a plane of one square centimetre area placed perpendicular to the direction of flow, where the pressure dif-

ference is one dyne per square centimetre of distance in the direction of flow.

- **capillary constant**
in a cylindrical tube of radius R, a liquid of capillary constant A, and of specific weight P rises by a height h such that: h=2A/Rp.

- **capillary fringe**
the ground immediately above the water table which contains water held by capillary rise. The lower part of the fringe is fully saturated. The fringe rises higher with increasing fineness of the soil and as the water table rises. It is the lowest part of the zone of aeration.

- **capillary fringe belt**
that portion of the aeration zone which lies between the intermediate belt and the water table. It contains fringe water.

- **capillary head**
the head in a capillary tube at any time prior to the time when the water rises to its ultimate position in the tube due to capillary force. It is equal to the difference in elevation between the ultimate position of the meniscus and the position of the meniscus at the given time. The head is zero when the water has risen to its ultimate position.

- **capillary lift**
the height to which water or other liquid rises in a capillary tube.

- **capillary migration**
movement of water through a rock or soil produced by molecular attraction of the rock or soil material for the water. It occurs in the zone of aeration and is most pronounced in those parts of the capillary fringe that come in contact with the belt of soil water, where there is a constant removal of water at the top of the fringe by evaporation or capillary fringe comes in contact with the zone of saturation.

- **capillary opening**
an opening small enough in cross-sectional area to create a condition of capillarity.

- **capillary percolation**
percolation of water due to capillary attraction through capillary interstices.

- **capillary retention**
the retention of water in the small pores by the soil. It evaporates slowly.

- **capillary tube**
a tube with an interior area of such small size that water

within it is raised or moved by capillarity.

- **capillary water**
  is the water held by surface tension in soil capillaries against the pull of gravity after the free or gravitational water has drained away. It is held between 31 and 1/3rd atmosphere. It moves from a point of saturation into drier parts of the soil.

- **capillary zone**
  the zone in which soil water is held by capillary forces the interstices may be completely filled (saturation zone) or partly filled (aeration zone). Pressure within this zone is less than atmospheric.

- **capital**
  all the durable and non-durable items used in production.

- **capital coefficient**
  also known as capital output ratio. It is the ratio between the book value of capital (machinery and equipment) in a plant and the gross value of its output. The capital-output ratio is the amount of capital necessary to produce an additional unit of output. It varies from one industry to another.

- **capital formation**
  it is also known as capital accumulation. It means increasing a country's stock of real capital. It implies additions to the existing supply of capital goods in a country. It represents the addition of new capital stock to existing stock after deducting depreciation, damage and other physical deterioration of the existing capital stock. Economic progress in a country depends upon its rate of capital formation.

- **capital intensive industry**
  it refers to that industry which uses large amounts of capital equipment in relation to its labour force or its output.

- **capital intensive technique**
  the method of production which uses a higher proportion of capital relative to other factors of production.

- **capital marginal efficiency of (MEC)**
  the relation between the prospective yield of one more unit

of a type of capital asset and the cost of producing that unit. The MEC of a particular asset means what an investor expects to earn from an additional unit of it compared with it costs him.

- **capitalist**
a plant which has stored food reserves.

- **capture**
a process involving a collision between a particle and a nucleus resulting in the retention of the particle within the nucleus and the emission of electromagnetic radiation called capture gamma rays.

- **capture gamma rays**
a nuclear capture process that results in the emission of gamma rays only.

- **caramel**
when sugar is heated and it gets darken, it is called caramel.

- **carbohydrate gum**
a polysaccharide which produces a gel of a viscous solution when it is dispersed in water at low concentrations.

- **carbohydrate nitrogen ratio (c:n ratio)**
the relative proportion of accumulated carbohydrates and nitrogen in stems and leaves of plants, changes with age, vigour, fruit-setting, etc., of the plant.

- **carbon 14**
a radioactive isotope that can be used for dating. This isotope is especially useful in dating material that was once alive, because all living matter contains carbon. The Tialf-life of carbon 14 is 5570±30 years. A radioactive isotope of carbon with atomic weight 14, produced by collisions between neutrons and atmospheric nitrogen. Useful in determining the age of carbonaceous material younger than 30,000 years.

- **carbon assimilation**
photosynthesis results in the incorporation of carbon dioxide into organic carbon.

- **carbon bonded sulphur**
that fraction of soil sulphur which is reduced to $S^{2\prime}$ by Raney nickel. This catalyst will reduce cystine and methionine but not ester sulphate. A close correlation has been reported between the concentration of Raney Ni-reducible S and the sum of

the concentrations of these two amino acids.

- **carbon cycle**

the sequence of transformations whereby carbon dioxide is fixed in living organisms by photosynthesis or by chemosynthesis, liberated by respiration and by the death and decomposition of the fixing organism, used by heterotropic species, and ultimately returned to its original state to be reused by another organism.

- **carbon nitrogen ratio**

the ratio of the weight of total organic carbon to the weight of total nitrogen in a soil or in an organic material. The C:N ratio of wheat straw is nearly 70:1 while that of soil is 10:1. When undecomposed straw with a high C:N ratio is applied to the soil, its C:N ratio is reduced through bacterial decomposition. To speed up bacterial decomposition, nitrogenous fertiliser is added to the soil at the time of turning under straw.

- **carbonate accumulation, soil horizon**

a developed soil horizon, beneath the surface, containing more calcium (or magnesium and calcium carbonate) than the soil above it or the material below it.

- **carbonation**

chemical weathering of rocks by $CO_2$ containing rain water especially on limestones and on rocks containing other basic oxides. The dissolved $CO_2$ functions as weak carbonic acid ($H_2CO_3$) which dissolves limestone as calcium bicarbonate.

- **carrier hypothesis**

according to Epstein and Hagen (1952) there is a carrier present on the membrane which combines with the ion at the outer surface of the plasmalemma, the carrier-ion complex so formed can traverse through the membrane, which is relatively impermeable to free ion. On releasing the ion, the carrier moves back to the outer surface of the membrane where it can again combine with an ion. It is visualised that a number of carriers are present in plants. The carrier hypothesis can explain

1. the accumulation of salts against the concentration gradient.
2. the specificity in ion uptake which is found in all plants

3. the transport of ions through membranes which by themselves are impermeable.

- **carrier of disease**
  a person or animal that harbours the organisms of disease without showing symptoms.

- **carryover effects (fertilisers)**
  the amount of nutrients left in the soil following a crop to which they have been applied. The amount remaining will depend on the amounts added, the yield, the method of harvesting, and the soil effects. What is left, however, is not all carryover because of leaching losses, fixation, and surface erosion.

- **carryover soil moisture**
  the moisture stored in crop root zone between cropping seasons or before the crop is planted. Moisture thus retained in the soil profile contributes to the water needs of crops before the irrigation are started. This contribution depends on the capacity of different soils to retain and supply moisture to crop plants. The balance between soil moisture at sowing and at harvest determines the actual amount of soil moisture contribution towards the water requirement of the crops.

- **cart**

  a common and popular means of rural transport and is widely used by farmers for the carriage of fodder and grain from the farm to home. Produce is also carried to the market.

- **cartel**
  it is a monopolistic organisation established for the purpose of restricting the output of member-firms in order to keep up the price of their product. The cartels first made their appearance in Germany.

- **cartograms**
  a type of diagrammatic and graphical representation of data where in maps are drawn and the figures representing

the phenomena at various places are shown by signs or symbols.

- **cartographic generalisation**
the practice of omitting boundaries between soil bodies as shown on detailed large-scale soil maps to reduce the detail on map- or between different taxonomic units if taxa of levels higher than the soil series are used to define and describe the basic map units. That is, it results in a lowering of the number of soil delineation.

- **cartography**
the preparation of the soil map in colour with appropriate legend depicting the various soil characteristics.

- **caryophyllatus**
it is a Latin word that means 'having a log claw'.

- **cash crops**
the crops produced entirely for the market or commercial purposes for e.g., tea, rubber, coffee, cocoa etc.

- **cash rent**
the system involving payment of fixed cash rent for the use of farm land or equipment.

- **cash share rent**
when share as well as cash rent is paid for the land and equipment used.

- **casual**
1. an occasional weed of cultivation.
2. a plant occurring in a plant community of which it is not a regular habitat.

- **casual labour**
labour employed on daily wage basis. The wages may be paid either in cash or kind or both.

- **cat clay**
set clay soils high in reduced forms of sulphur which upon being drained become extremely acid due to the oxidation of sulphur compounds.

- **catabolism**
metabolic reactions involved in the breakdown of complex molecules to simple compounds. The function of catabolic reactions is to provide energy, which is used in the synthesis of new structures, for work (for e.g., contraction of muscles), for transmission

of nerve impulses, and for the maintenance of functional efficiency.

- **catalase**
an enzyme that breaks down hydrogen peroxide (which is toxic) to water and oxygen. During this reaction it is slowly destroyed itself. Its prosthetic group contains iron, and its maximum activity is between pH 6.5 and 7.0. It is present in all aerobic tissues.

- **catalyst**
substance whose presence augments the rate of a-chemical reaction and it (substance) may be recovered essentially unaltered in form and amount at the end of the reaction.

- **cataphoresis (electropheresis)**
the migration of colloidal particles in an electric field, especially towards the cathode.

- **catch crop**
1. a crop grown temporarily on land that will eventually carry a crop of different species, in order to maintain the productivity or prevent the erosion of the land prior to the-establishment of the main crop.
2. a crop employed to stimulate germination and growth of some specific, parasitic weed and then destroyed before planting the main crop.

- **catchment**
the area from which a single river system collects its waters. The boundary of a catchment area is defined by the watershed.

- **catchment area**
the total area draining into a given waterway, lake or reservoir.

- **catchment water harvesting**
the runoff water is collected from a treated or untreated catchment and stored in a reservoir or farm pond. This stored water is utilised for supplemental irrigation during long dry spells and at critical stages of plant growth.

- **categorical generalisation**
the identification and description of soil map units in terms of taxa at levels of abstraction higher than the soil series. That is, it involves use of taxonomic unit name and thus reduce the classification of taxonomic detail.

- **category (soil taxonomy)**
a series or array of taxa produced by differentiation

within the population at a given level of abstraction or generalisation, and is composed of all the classes at one level of generalisation.

- **catena**
  a Latin word that means 'chain'. A soil catena is usually associated with undulating landscape. Quite different soils may be developed at the top, middle, and bottom of the slopes, though the parent material may be the same. Such soils are associated like the links of a chain by the topography and are mapped as catena.

- **cation**
  a lump of earth which holds together, larger than a clod. It indicates a relatively clayey soil.

- **cation analysis**
  qualitative analysis for cations in aqueous solution.

- **cation antagonism**
  exceptionally strong competition between pairs of similar cations. The specific nature of antagonism indicates that the ions involved enter the root the same way whereas other ions may have other entries. Most plant physiologists postulate the existence of several kinds of carrier molecules that selectively transport ions into the inner space of roots. Similar ions must enter by the same carrier molecules. They are therefore so strongly competitive for entry that they are antagonistic to each other. Pairs of similar anions likewise exhibit antagonism and are believed to share carrier molecule.

- **cation exchange capacity**
  the milliequivalents of cations per 100 gm, of soil which can be held by surface forces and which can be replaced by other cations. The term as applied to soils is synonymous with base exchange capacity but is somewhat more precise in its meaning. Cation exchange capacity depends somewhat on the method used for its determination. For example, using a different replacing cation or making the replacement at a different pH affects the result. For western soils it is usually determined by leaching with neutral ammonium acetate and determining the absorbed ammonia.

- **cation fixation**
  is the reduction of exchangeability or extractability of cations due to interactions with the soil's solid phase.

**cation substitution**
a substitution involving cations which fit into the mineral structure with no change in external form. It occurs within the layers of all 2:1 layer silicates. One negative charge from the surrounding O— ions is left over wherever an $Mg^{+4}$ or an $Fe^{++}$ ion replaces an $Al^{+++}$ ion in an octahedral sheet or an $Al^{+++}$ replaces an $Si^{++++}$ ion in a tetrahedral sheet. These negative charges are balanced by cations held between or outside the layers.

**cattle yard report**
day to day events happening in cattle yard are recorded which may include birth and death or treatment of sick animals as well as special wants like inoculation and other treatments.

**caul**
Latin for 'stem'.

**cauline**
belonging to the stem or arising from it.

**caulis**
a Latin prefix meaning 'stemmed'.

**caulo**
a prefix from the Greek meaning 'stem'.

**caulocarpic**
said of a plant, which, after flowering survives for the rest of the year, to flower in the next or subsequent years.

**causal organism**
an organism such as fungus, bacterium, nematode, etc., which produces a particular disease.

**caustic**
an agent capable of corroding or eating away tissues, burning, corrosive.

**caution**
a warning to the user of pesticide chemicals. Used on labels of pesticide containers having slightly toxic pesticides in toxicity as denned by the Insecticide, Fungicide and Rodenticide Act.

**cavity well**
a type of tube well where water is directly pumped from the cavity created at the bottom of the bore. These are considered as cheapest among the tube wells.

**ceiling**
the upper limit of the highest class in a series of classes.

**ceiling of independence principles (soil taxonomy)**
a property or characteristic used as a differentiating characteristic in a category

must not separate similar individuals in a lower category. Every characteristic is limited in its use by a ceiling categorical level above which it cannot be applied without injecting confusing and inappropriate cleavages in a lower categories.

- **ceiling on land holdings**
the law passed to fix ceilings on agricultural land holdings have been enacted by a few States and is under consideration in other states of the country. Ceiling on land holdings is intended to reduce the disparities in incomes and promote the development of the rural economy. The ceiling when enforced is expected to make available the surplus land for distribution among the landless persons in the agricultural sector.

- **ceiling price**
it is the maximum price laid down by the government under a system of price control. Ceiling prices are generally imposed in wartime, though they have also been fixed in peacetime as an anti-inflationary measure.

- **cell**
the fundamental unit of life, the smallest body capable of independent reproduction. Cells are always surrounded by a membrane.

- **cell body**
the part of the cell containing nucleus and some cytoplasm.

- **cell cycle**
it refers to the sequence of events in dividing cells including the gi, S, 62 and M periods.

- **cell differentiation**
the process whereby descendants of a common parental cell achieve and maintain specialisation of structure and function.

- **cell division**
it is the process of formation of two daughter cells from a parent cell by enclosure of the two nuclei in separate cell compartments.

- **cell exchange**
exchanges of chemical substances between the cytoplasm and the internal environment. These exchanges allow for the elimination of waste matter of metabolism and the absorption of the substrates necessary for cell permeability.

- **cell free extract**
a fluid containing most of the soluble molecules of a cell, made by breaking open cells and getting rid of remaining whole cells.

- **cell fusion**

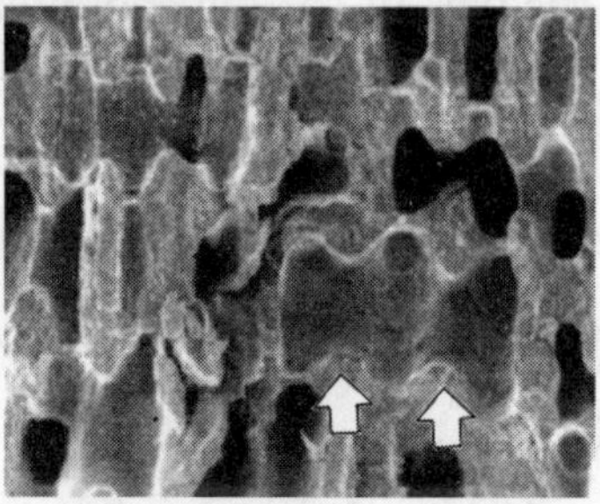

formation of a single hybrid cell with nuclei and cytoplasm from different cells. Often induced by treatment of mixed cell culture with dead sender virus.

- **cell membrane (plasma membrane)**
the external lining of the protoplasm constitutes the outer boundary of animal cells which lack cell walls. Chemically lipoprotein. The lipid forms the fluid in which proteins, extrinsic (peripheral) and intrinsic (integral) are distributed and contribute a mosaic pattern. Membrane appeared triple-layered in the electron micrograph, two dark bands separated by a light band. Has a total thickness of 75 A° to 100 A°. Membrane is responsible for differential permeability.

- **cell plate**
plate of materials appearing across the equatorial plane of the spindle during telophase in plant cells. Develops in middle lamella.

- **cell sap**
a solution of organic and inorganic substances in the vacuole of a plant cell.

- **cell theory**
the theory that all living things are made up of cells that divide during growth and reproduction.

- **cell wall**
the non-living rigid structure surrounding the cell membrane of most plant cells, usually made of cellulose in green plants, sometimes with other materials such as lignin or cutin, made of chitin in some fungi, and of various materials in prokaryotes.

- **cellular affinity**
tendency of cells to adhere specifically to cells of same type,

but not of different types. This property is lost in cancer cells.

- **cellulase**

an enzyme capable of digesting cellulose into simpler units. Associated with sapropytes and parasites, and particularly important which when produced by the micro-flora in the herbivore's large intestine.

- **cellulolytic power of the soil**

the biological activity of the soil can be measured by the decrease in the weight of a circle of filter paper which is digested by the bacteria.

- **cellulose**

a polymer of glucose. The actual breakdown of cellulose and the resultant products depends on the nature of the organisms and the conditions of decomposition. The aerobic bacteria and fungi break down cellulose, completely, producing only carbon dioxide, and a large amount of microbial cell-substances. The anaerobic bacteria, however, decompose the cellulose with the formation of various organic acids-and alcohol as shown below:

$(C_6H_{10}0_5)-^{+} (n-1) H_20=n C_6H_{12}O_6$

$C_6H_{12}O_6 = C_3H_60_3 + C_2H_50H+CO_2$ $2C_2H_5OH=CH_3COOH + 2CH_4$

The dung of animals contain under composed portions of cellulose and the degradation products of cellulose as the animal tract harbours cellulose, decomposing bacteria.

- **celsius scale**

a temperature scale propounded in 1742 by Anders Celsius in which the temperature of melting pure ice is taken as 0° and the temperature of boiling water 100° (both at standard pressure), The degree Celsius (°C) is equal to the Kelvin.

- **cement kiln potassium**

in the production of cement some of the potassium bearing ingredients are given off in flue dusts-and gases. These dusts and fumes are collected in a precipitator, and the product is sold as a fertiliser for its potassium and calcium content. Such material contains about 30 per cent calcium and 5 per cent potassium.

- **cementation**

cementation of soil material refers to a brittle hard consistence caused by some cementing substance other than clay minerals, such as calcium carbonate, silica, or oxides or salts of iron and aluminium. Typically, the cementation is altered little if any by moistening: the hardness and brittleness persist in the wet condition. Cementation may be either continuous or discontinuous within a given horizon.

a) Weakly Cemented: Cemented mass is brittle and hard but can be broken in the hands.

b) Strongly Cemented: Cemented mass is brittle and harder than can be broken in the hand but is easily broken with a hammer.

c) Indurated: Very strongly cemented, brittle, does not soften if wetted for a long time, so hard that a strong blow with hammer is required to break, hammer generally rings as a result of the blow.

- **centi**

a prefix denoting $10^2$ for example, 1 centimetre (cm)=$10^2$ metre (m).

- **centigrade (°c)**

a thermometer scale in which water freezes at 0 degrees C. and boils at 100 degrees C. To change to degrees Fahrenheit, multiply degrees centigrade by nine-fifths and add 32.

- **central dogma**

the basic relationship between DNA, RNA, and protein: DNA serves as a template for both its own duplication and the synthesis of RNA, and RNA, in turn is the template in protein synthesis.

- **central planning**

it refers to that system of economic planning where the State determines what shall be produced, how much shall be produced and how shall it be produced. The State allocates factors of production to the various industries for productive purposes. Communism and central planning often go hand in hand. Central planning is now increasing even in non-communist societies.

- **central tendency**

the tendency of the observations or cases in a distribution to cluster about a point, with respect either to absolute value or to frequency of occurrence, usually but not necessarily about midway between the extreme high and extreme low values in the distribution.

- **centralised planning**

the type of planning in which the allocation of resources is mainly done by the central planning agencies.

- **centrifugal**

1. a principle of force, moving or flying away from the centre.

2. in Botany, growing out-

ward from the centre.

- **centrifugal force**
a force caused by rotation which causes motion away or out from the rotating object. Serves to cause solids or liquids to separate from liquids of lower specific gravity.

- **centrifugal separation**
the separation of substances of different densities by means of centrifugal force, as the separation of fat from milk.

- **centrifuge**
a device for separating the components of a suspension by spinning. In simple types, the liquid containing suspended particles is placed in a tube, which is whirled round in a horizontal circle. Particles less dense than the liquid move to the top of the tubs, denser particles accumulate at the bottom.

- **centripetal**
towards the centre.

- **centromere**
it refers to the primary constriction of the chromosome to which the spindle fibres are attached and which helps in chromosome movement to the poles at anaphase.

- **centroplasm**
a mass of protoplasm in the centre of the cells of one of the Myxophyceae. Its function is obscure.

- **cepillary rise**
the rise of water above the phreatic surface through the influence of capillarity.

- **cerat**
Greek prefix meaning 'horn'.

- **cereal forage**
a cereal crop harvested when immature for either hay, silage, green feed or as pasturage.

- **cereal or grain crops**
cereals are grasses grown for their edible seeds, the term cereal being applied either to the grain or to the plant itself. They include wheat, oats, barley, rye, rice, maize, sorghum, millet, etc. Grain is a collective term applied to cereals.

- **cernuous, ceruus**
Latin meaning 'hanging, drooping, nodding'.

▪ **certified seed**
the progeny of foundation, registered, or certified seed that is so handled as to maintain satisfactory genetic identity and purity and that has been approved and certified by the certifying agency as conforming to certain standards of pedigree and excellence, and free from disease. This seed is made available to farmers on mass scale.

▪ **ceteris paribus**
it is a Latin phrase for 'other things being equal'. It is one of the frequent assumptions made in economic theory when we study a particular aspect of a problem with all disturbing elements removed.

▪ **cf horizon**
a C-horizon, which is disintegrated but not chemically transformed.

▪ **cf m per bu**
used for specifying air flow in cu ft per min. for each bushel of crop, obtained by dividing the Cfm of air by the bin capacity in bushels where the capacity of the bin is determined on the basis of 1.25 cu ft for 1 bu for small grains, and 2.50 cu ft for 1 bu for clean husked ear corn, employed in connection with rate of air flow for conditioning of harvested produce.

▪ **CF S**
cubic feet per second of flow of water in a stream.

▪ **cfm per cu ft**
used for specifying air flow in cu ft per min. for each cu ft of crop, obtained by dividing the Cfm of air by total bin volume in cu ft., employed in connection with rate of air flow for conditioning of harvested produce.

▪ **cfm per sq ft**
1. used for specifying air flow in cu ft per min. on the basis of a floor area of 1 sq ft, obtained by dividing the cfm of air by the area of the floor in sq. ft, employed in connection with rate of air flow for conditioning of harvested produce.

2. u sed for specifying airflow through forages, bulk root crops, etc., where the total air flow in cfm is divided by tons of product in storage. The capacity in tons is determined on the basis of volume occupied by crops at the moisture content normally stored, employed in connection with rate

of air flow for conditioning of harvested produce.

- **chacre**
a kind of calcium carbonate found at a shallow depth in the subsoil and used as an amendment.

- **chaff**
1. outer covering of grain, the glumes, the husk, that is removed by threshing.
2. finely cut hay or straw or green stalks used as feed for cattle.

- **chain harrow**
a harrow having no rigid frame, but consisting of interlocked, round, square, or different shaped links which make up a strong, flexible mesh to adjust to the inequalities of the ground surface. Used for rolling up field and collecting weeds uprooted by the working of other implements.

- **chain reaction**
a chemical reaction in which many molecules undergo chemical reaction after one molecule becomes activated.

- **chaining**
bruising of rooted aquatic weeds with a heavy chain in ditches filled with water.

- **chak**
plot of land as shown in the land records

- **chamae**
a Greek prefix meaning 'ground'.

- **chance seedling**
a fruit or nut tree which grows from seed scattered by wind, birds, etc, sometimes bears fruit. Many varieties of fruits originated as chance seedlings.

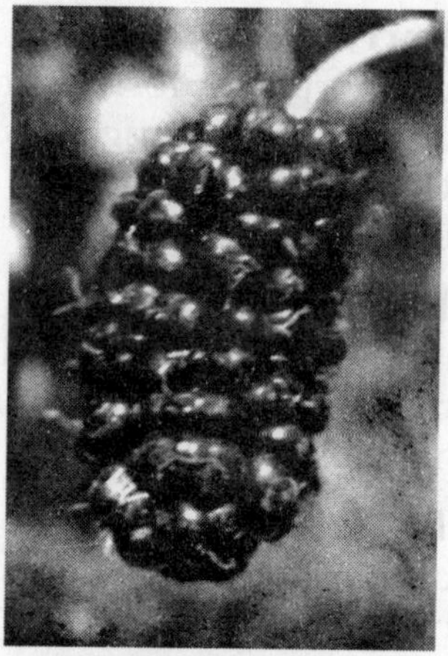

- **change agent**
1. one who advocates or attempts to introduce another method, a teacher, an agricultural extension worker
2. the new system itself.

- **channery**
  soils contain fragments of thin, fiat sandstone, limestone, or schist up to 6 inches along the longer axis. A single piece is a fragment.

- **char filters**
  a term used in sugar refinery which means decolouring sugar solution by bone char or vegetable carbons.

- **character (characteristic)**
  a distinguishing, feature, trait or property of an organism.

- **charcoal**
  an amorphous form of carbon made by heating wood or other organic material in the absence of air.

- **chase**
  a tract of unenclosed land reserved for breeding and hunting wild animals.

- **chasmogamic**
  said of a flower that opens before pollination.

- **chasmophyte**
  a plant growing on rocks, rooted in debris in the crevices.

- **check basin (irrigation)**
  irrigation applied to relatively level plots bonded on all sides. The basin is a small check.

- **check dam**
  a small low dam constructed in a gully or other watercourse to decrease the velocity of stream flow and thereby to minimise channel scour and promote deposition of eroded material.

- **check gates**
  are gates placed in an irrigation channel to form an adjustable dam to control the elevation of the water surface upstream.

- **check irrigation**
  a method of irrigation where the water is applied to a field or orchard which has been divided into a series of level checks by earth ridges, the water flowing from one check to another along the slope. The checks may be 1/4 hectare to 1 hectare in size.

- **check plot**
  a small area which is used as

a standard for comparison in an agricultural experiment.

- **check row**
  a row of seeds or plants spaced equally in both directions.

- **check row planting**
  a method of planting in which row to row and plant to plant distance is uniform. In this method, seeds are planted precisely along straight parallel furrows. The rows are always in two perpendicular directions.

- **cheek valve**
  a valve provided with a disk hinged on one edge so that it opens in the direction of normal flow, and closes with reversal of flow.

- **chelates**
  1. the term chelate is derived from the Greek word meaning claw.
  2. water-soluble organic compounds that can immobilise metallic cations.

- **chemical**
  1. related to the science of chemistry.
  2. substance characterised by definite molecular composition.

- **chemical eluviations**
  eluviations in which decomposition occurs and certain products liberated are removed in true, colloidal solution to be deposited in other horizons.

- **chemical fertiliser**
  any chemical compound used for supplying one or more of the essential plant food elements.

- **chemical inhibitor**
  a substance capable of stopping or retarding chemical reaction.

- **chemical potential**
  is defined as, if unit quantity of that substance passes from one potential to another, the difference in potential is equal to the maximum - amount of work involved.

- **chemical potential of water**
  the chemical potential of a substance in a system is a measure of the capacity of that substance to work. In a simple solution of a non-electrolyte in water, the chemical potential of water depends on the mean free energy per molecule and the concentration of water molecule.

- **chemical reactivity**
the tendency of two or more chemicals to react to form one or more products differing from the reactants.

- **chemical weed control**
the control of weeds in crops by the application of herbicides as pre-emergence and post-emergence sprays.

- **chemically inactive**
refer to that chemical or substance that will not easily react with any other chemical or object.

- **chemically pure**
without impurities detectable by analysis and abbreviation CP.

- **chemisorbed phosphorus**
phosphorus adsorbed or precipitated on the surface of clay minerals or other crystalline materials as a result of the attractive forces between the phosphate ion and constituents in the surface of the solid phase.

- **chemisorption**
adsorption usually involving primary valence bonds and therefore stronger and more specific than physical adsorption.

- **chemoautotrophic**
obtaining energy from a simple inorganic reaction, the nature of which varies according to species. Several kinds of autotrophic bacteria are chemoauto-trophic, for e.g., Thiobacilli.

- **chemolysis**
decomposition of organic matter brought about by chemical agents.

- **chemosynthesis (chemosynthetic)**
a method by which micro organisms, including chemotrophs, obtain energy. They use the chemical energy of inorganic or organic materials (but not photosynthesis) to build up their organic matter.

- **chemotherapy**
the prevention or control of infectious disease in man, animals or plants by the use of chemical agents.

- **chemotrophic**
of organisms, obtaining energy by chemical reactions independent of light. The reactions may be based on inorganic substances or organic

substances obtained from the environment.

■ **chemotropism (BOT)**
tropism in which stimulus has the form of a gradient of chemical concentration, for e.g., downward growth of pollen tubes into stigma due to presence of sugars.

■ **chena**
a burning of forest for a few crops.

■ **chernozem soils**
1. a class of soil in Gedroiz (1927) classification which is based on the nature of the cations absorbed by the organic and inorganic soil colloids.
2. chernozem soils have predominance of calcium in the absorbing complex. These soils are deep, dark coloured to nearly black surface horizon, rich in organic matter grades below into lighter coloured soil and finally into a layer of lime accumulation, temperate to cool subhumid climate. Name derived from the Russian for black earth.

■ **chersophyte**
a plant which grows on wasteland or on shallow soil.

■ **cherty soils**
soils having angular fragments that are less than 3 inches in diameter, more than 75 per cent of which are chert, coarse cherty soils have fragments of 3 to 10 inches. Unless the size distinction is significant to the use capability of the soil, the cherty soil includes the whole range up to 10 inches. Most cherty soils are developed from weathered cherty limestone.

■ **chestnut soil**
a grassland soil developed under arid conditions. There is no leaching, and it carries shallow rooting vegetation. The calcium carbonate layer is near the surface. They occur on the arid side of chernozem soils, into which they grade.

■ **chezy formula**
1. an empirical formula for the velocity of a uniform flow

of water through a section of a stream in terms of the conduit roughness, the hydraulic radius and the friction slope. The formula is usually written: V=C √RS.

2. The Chezy formula derived its name from Antoine de Chezy, a celebrated French engineer of the eighteenth century.

- **chi square ($x^2$)**

a statistic used to test the agreement of obtained and theoretical distribution, mathematically equal to the sum of the quotients obtained by dividing the square of each difference between the actual and the theoretical frequency (for each category) by the theoretical frequency for that category.

- **chi square test of goodness of fit**

a statistical procedure by means of which the probability can be estimated that a given fit of data or one showing a greater discrepancy might arise if a certain law or cause were in operation based on. the following formula: Where $f_0$ is each observed frequency and ft is each theoretical frequency, determined by calculating $X^2$ and then finding the probability that a Chi square of that or greater magnitude might occur by chance alone by reference to a table of the sampling distribution of $X^2$.

- **chiasma**

the attachment of two non-sister chromatids in a bivalent that is first seen in diplotene of prophase I of meiosis. Each chiasma results in the exchange of genetic material between the non-sister chromatids, *i.e.*, in crossing over. The plural is chiasmata.

- **chill**

1. sensation of coldness, attended by shivering, may be followed by a variety of other disease symptoms in animals.
2. to reduce the temperature of a product and increase its life and value.

- **chimaera**

a plant having tissues of more than one genetic type. This may be caused by the mutation or abnormal distribution of chromosomes in one cell early in development, thus affecting tissues derived from it. Grafting different types of plants—a graft hybrid, can produce it.

- **chipping**
hand hoeing amongst the crop stools to remove weeds, grass or vines which cannot be effectively controlled mechanically or chemically.

- **chisel**
a tillage implement with a strong narrow cultivator-type used to shatter or loosen hard, compact layers, usually in the sub-soil at depths below normal plough depth.

- **chisel ploughing**
cultivating with a heavy tyned implement.

- **chisel technology**
the technology involves chiselling of the dry soil up to 40 cm depth at 50-120 cm intervals, depending upon the row to row spacing of the seed lines. It has been developed for the management of soils having layers of high mechanical impedance at shallow depths such as plough soles or pans of clay accumulation in B-horizon.

- **chisel tillage**
a tillage practice in which narrow chisel 5 to 10 cm wide is run through the soil at a depth of 15 to 30 cm leaving all plant residues on the surface. The chisel shatters tillage pans and surface crusts. Planting is done through the residue with special planter.

- **chiselling**
the breaking or shattering of compact soil or subsoil layers by the use of a chisel.

- **chlamydomorphous soil**
a soil in which each grain has a colloidal coating and the grains are bound together in a loose aggregate.

- **chloride accumulation in soil**
since citrus is highly sensitive to chloride content of waters, a classification has been set up in Israel based on large scale survey of irrigation waters used for citrus orchard of that country. They have worked out an equation to predict the chloride accumulation in soils, which is:
Where, Clw=chloride content of irrigation water, SP=saturation percentage of the soil, Cls=Chloride accumulation in soil, R=Mean annual rainfall.
Chloride accumulation increases with chloride content of irrigation water and the saturation percentage value of the soil.

- **chloride percentage**
the ratio of the total chloride anions or the total chloride plus nitrate anions, to the total anions held in solution, multiplied by 100.

- **chloride toxicity symptoms**
it results in reduction in the number and the size of the leaves, the bronzing or yellowing of the foliage and the browning and scorching of the leaf margins.

- **chlorinator**
a device for adding chlorine gas to sewage to kill germs.

- **chlorine**
an essential micro-nutrient for plant growth. Most of the chlorine deficiencies have been obtained with nutrient culture studies in the green house. Chlorine has been shown to be involved in the oxygen evolution in primary photosynthetic reactions. Plants responding to this element are: tomatoes, peas, lettuce, cabbage, carrots, sugar beets, barley, corn and cotton. However, in excessive quantities it has a detrimental effect on some plants for e.g., as of tobacco and potatoes. In both species, the leaves become thickened and tend to roll.

- **chlorine deficiency symptoms**
so far observed only in plants grown at low, deficient level of the nutrient in rained sand or water-culture. The symptoms being the wilted appearance of the foliage and stuffy roots with the laterals showing branching. In some plant species for e.g., tomato leaves also show chlorotic mottling, bronzing and tissue necrosis.

- **chlorophyll**
green pigment found in plants, which absorbs energy from sunlight, enabling them to build-up carbohydrates from atmospheric $CO_2$ and $H_2O$ by photosynthesis.

- **chlorophyll a**
a blue-green pigment comprising 70% of chlorophyll. It is the magnesium complex of the phytyl ester of a porphyriu-propionic acid, $C_55H_{72}OMgN_4O_5$. The alcoholic solution has a blue-green colour with a red fluorescence.

- **chlorophyll b**
a yellow-green colour comprising 30% of chlorophyll. It is the magnesium complex of

the phytyl ester of a porphyrin-propionic acid, $C_55H_{70}MgN_4O_6$. The alcoholic solution has a yellow-green colour with a red fluorescence.

- **chlorophyllase**

an enzyme concerned with the synthesis or breakdown of the chlorophyll molecule. It functions in the attachment or removal of the phytol side-chains to the porphyrin nucleus of the chlorophyll, and prochlorophyll units.

- **chloroplastid**

a special body in a cell, generally a plant cell, which contains chlorophyll.

- **chlorosis**

1. a condition in plants resulting from the failure of chlorophyll, develop caused by a deficiency of an essential nutrient. Leaves of chlorotic plants range from light green through yellow to almost white.

2. It is a deficiency symptom due to inadequacy or deficiency of minerals like nitrogen, magnesium, iron, etc. It may also be caused by lack of light which prevents chlorophyll formation.

- **chlorothion(0, 0-dimethyl 0, 4-nitro-3-chlorophenyi-thiophosphate)**

is one of the two chlorinated analogues of methyl parathion with more reduced mammalian toxicity.

- **choke weed**

refers to those weeds that crowd out other plants

- **chops**

a mixture of dehydrated leaves and stems as they come from the dehydrator.

- **chresard**

the total amount of water in the soil which is available to plants.

- **chroma**

a basic component of colour which defines the gradation of purity of colour or the apparent degree of departure from neutral greys or white.

- **chromatic adaptation**

variation in the colouration in relation to the amount of light reaching a plant.

- **chromatid**
  a half chromosome during duplication in early prophase and metaphase of mitosis and between diplotene and the second metaphase of meiosis. After these stages it is called a daughter chromosome.

- **chromatin**
  a nucleoprotein, staining with basic dyes, and forming part of the chromosomes. It is made up of two proteins one of which is histone DNA and RNA.

- **chromatogram**
  the pattern formed by zones of separated pigments and of colourless substance in chromatographic procedures.

- **chromatographic adsorption**
  preferential adsorption of chemical compounds (gases or liquids) in an ascending molecular-weight sequence onto a solid adsorbent material, such as activated carbon, alumina or silica gel, used for analysis and separation of chemical mixtures.

- **chromatography**
  a method of chemical analysis in which a mobile phase carrying the mixture to be analysed, is caused to move in contact with a selectively absorbent stationary phase. The mobile phase may be a solution of a mixture of compounds in a suitable inert solvent or it may be a mixture of compounds in a vapour diluted with an inert carrier gas. The stationary phase may be an absorbent (active) solid or liquid supported on an absorbent solid, it is characterised by its ability to retain the components of the mixture to different degrees. During the progress of the mobile phase in contact with the stationary phase, the components of the mixture become separated and can be identified, in some cases they can be determined quantitatively. When the mobile phase is a gas and the stationary phase is a liquid on a solid support the process is known as gas liquid.

- **chromatoplasm**
  in blue-green algae, the peripheral contents of the cell, containing the photosynthetic pigments.

- **chromiole**
  one of the minute granules of which a chromo-mere is composed.

- **chromium (dichromate) toxity symptoms**
  plants show a spreading habit. In cereals, the leaf sheaths and

the upper part of the stem often turns brown. Old leaves show scorching, first near the leaf margins and later all over the lamina. The lamina may also develop purple colouration.

- **chromomeres**
darkly staining bead-like granules found in constant arrangement along prophase chromosomes during meiosis. Chromomeres in corresponding positions on homologous chromosomes pair during meiosis in many organisms. Probably mark positions where chromosome thread is more tightly coiled than in other regions.

- **chromoparous**
said of colourless bacteria which secretes a colourless material.

- **chromophorous**
of bacteria having colouring materials as part of their make up.

- **chromoplast**
a plastid containing one or more coloured pigments.

- **chromosome**
the physical basis of heredity. Deeply staining rod-like structures present with the nuclei of eukaryotic. Chemically nucleoproteins since they contain DNA and histones arranged in compact manner. Number is constant in cells of a particular species. Are replicated into identical chromatids during mitosis and meiosis. Chromatids separate and go to daughter cells as chromosomes.

- **chromosome arm**
one of the two parts of a chromosome to which the spindle fibre is attached along the side.

- **chromosome complement**
the set of chromosomes characteristics of any cell of a species.

- **chromosome map**
a diagram of a chromosome showing the position of the genes. It is obtained by a statistical analysis of the frequency of crossing-over.

- **chronic**
1. in the medical sense, this term is used in contrast to 'acute' and means 'of long duration'.
2. As applied to material inhaled or absorbed through the

skin, it means exposure times of long duration.
3. As regards materials ingested, it refers to repeated doses.

- **chronic effect**
slow and long continued effect.

- **chronomorphic soil**
when the appearance and development of soil are influenced primarily by the action of the weather, which is the permanent factor over long periods.

- **chronopedology**
the study of the processes of formation of soils and of their various evolutions, especially with respect to rates and durations, pedology considered as a function of the weather.

- **chuni**
a by-product of pulses containing husk and -broken pieces of grain, useful for feeding animals.

- **cipoletti weir**
a contracted trapezoidal weir in which each side of the notch has a slope of 1 horizontal to 4 vertical. The discharge through Cipoletti weir is computed by the following formula:
$Q=0.0186LH^{3}/^{2}$
Where, Q=discharge (litres/second) L=Length of crest (cm) H=Head over the crest (cm).

- **circular flow**
1. it means the continual circular movement of money and commodities in the economy.
2. It illustrates the flow of money and commodities from households to business enterprises and back to the households.

- **circular fodder cutter**
also called chaff cutter. Light implement consisting of a stand, a body and main wheel, body consisting of rollers, side plates to hold the rollers, shaft, gears, front plate, spring, and main wheel with knives.

- **circular paper chromatography**
paper chromatographic techniques in which migration from a spot in the sheet takes place in 360° so that zones separate as a series of concentric rings.

- **circulation**
1. the regular movement of any fluid within definite channels in the body.
2. the streaming movement of protoplasm of plant cells.

- **circumnutation**
autonomic growth movement in the growing tip of a

plant showing curvature in a helical manner around the whole circumference.

- **citrate soluble phosphorus**

the term is frequently encountered in describing the phosphate contained in fertilisers. The residue obtained from the determination of water-soluble phosphorus leaching process is added to a solution of neutral N ammonium citrate. It is extracted for a prescribed period by shaking, and the suspension is filtered. The phosphorus content of filtrate is determined, and the amount present, expressed as a percentage of total weight of the sample is termed the citrate soluble phosphorus.

- **clarification**

in waste water treatment, the removal of turbidity and suspended solids by settling, often aided by centrifugal action and chemically induced coagulation.

- **clarifiers**

1. in sugar manufacture, raw juice from mills or diffuser is limed heated and settled to separate clear juice from mud.
2. in waste water treatment, a setting tank that mechanically removes settled solids from wastes.

- **class**

used in soil taxonomy, a group of individual or other units similar in selected properties and distinguished from all other classes of the same population by differences in these properties.

- **classical liberalism**

1. it is also sometimes referred to as nineteenth century liberalism.
2. The philosophy of classical liberalism has strong faith in free and competitive markets unaccompanied by governmental restraints and controls. According to classical liberalism, human wants are best satisfied and productive resources most efficiently employed when commodities and services are sold in free competitive markets. This philosophy emphasises minimum role for governmental intervention.

- **classical school**

it refers to that school of thought among economists in the late 18th and early 19th centuries, which believed that national economies function better and more efficiently under private enterprise and vigorous competition than under government control.

The main differences between the Classical School and the Neo-classical School, which developed later, were the emphasis of the latter on mathematical economics and the psychological factors in the determination of consumer demands.

■ **classification**

is merely an arrangement or grouping of the objects based on such of their characteristics that will bring all of them into the smallest number of groups with a view to visualise the individual characteristics of the objects and study their interrelationships. The success of any system of classification is largely measured in terms of the precision with which the units are defined and named, the objects grouped under each and the universal acceptance accorded to it.

■ **classification soil**

the assignment of objects or units to groups within a system of categories distinguished by their properties. In the classification of soils, the fundamental unit is a soil type. Similar soil types are grouped to form a series. Series are grouped into families, families into Crest Soil Groups, these into suborders, and suborders into orders (of which there are three: the Zonal, Azonal, and Intrazonal).

■ **clastic**

compounds of broken fragments of rocks and minerals.

■ **clay**

1. soil particles less than 0.002 mm in size.

2. a complex group of minerals, composed of aluminium and iron silicates

3. soil material that contains 40 per cent or more clay, less than 45 per cent sand, and less than 40 per cent silt.

■ **clay complex**

consists of both clay and humus largely having similar properties, also called clay humus complex.

■ **clay cutan**

a kind of cutan largely composed of clay that has been transported by water through the larger voids in the soil and deposited on their walls at a time when the water left the voids. It is, therefore, a coating of clay. It is most easily identified where it is redder or darker coloured than the interior of the ped.

**clay humus complex**
a complex of particles of fine clay and humus, which has the property of attracting and holding ions dissociated from salts dissolved in soil moisture.

**clay loam**
soil material that contains 27 to 40 per cent clay and 20 to 45 per cent sand.

**clay mineral**
naturally occurring inorganic material (usually -crystalline) found in soils and other earthy deposits, the particles being of clay size *i.e.* less than 0.002 mm in diameter.

**clay pit**
a place from which clay is taken.

**clay pots**
these are familiar red clay flower pots, used for growing young plants. They are easily broken, and at times, their round shape is not suited as it takes additional space. They are however inexpensive and widely used.

**clay structure**
structure of clay particle, when viewed under the electron microscope, seen to be definitely crystalline and lathe-shaped or made up of irregular hexagonal slates.

**clayey**
includes all clay textural classes *i.e.* sandy clay, silty clay, and clay.

**claying**
incorporation of small amount of clay or heavy textured soil to light soils. It helps to bind the sand grains together and improves the water holding power of the soil.

**claypan**
a dense, compact layer in the sub-soil having a much higher clay content than the overlying material, from which it is separated by a sharply defined boundary, formed by downward movement of clay or by synthesis of clay in place during soil formation. Claypans are usually hard when dry, and plastic and sticky when wet. Also, they usually impede the movement of water and air, as well as the growth of plant roots.

**clean cultivation**
periodic soil tillage to elimi-

nate all vegetations other than the crop being grown.

- **clean rice**
milled rice free of hulls.

- **clean tillage**
means ploughing and cultivating to kill weeds and bury all crop residues. Such tillage is very expensive, causes soil compaction, induced pans, induced crusts and results in erosion. The basic aim of clean tillage is to prevent the growth of all vegetation except the particular crop desired.

- **cleaning and separation**
processes for removing the foreign materials like weed seeds, chaff, dead insects and broken stems from agricultural produce.

- **clear**
in land reclamation, to remove trees, bushes, stones or other obstacles to tillage.

- **clearing**
removing growth by bulldozing, chaining, rolling, logging, slashing, ploughing, raking, burning, or any combination of these.

- **clearing of stones**
the removal of stones from the soil to increase the percentage of fine earth and to facilitate the passage of implements.

- **cleavage mineral**
is the property of splitting or separating easily along planes which are definitely related to the interval arrangement of the molecules.

- **cleistogamy**
the condition of having flowers which never open and are self-pollinated, and are often small and inconspicuous.

- **climacteric**
the rise in respiration rate, found in some species, associated with fruit ripening and senescence.

- **climate**
1. the climate of a locality is the synthesis of' day-to-day values of meteorological elements that affect the locality. 2. the total long time characteristic weather of any region.

- **climate cycle**
actual or supposed recur-

rences of such weather phenomena as wet and dry years, hot and cold years, at more or less regular intervals, in response to long-range terrestrial and solar influences, such as volcanic dust and sunspots.

- **climate index**
an index number which condenses climatic data into a simplified expression.

- **climate oceanic**
the type of climate characteristic of land areas near oceans which contribute to the humidity and at the same time have a moderating influence on temperature and the range of temperature variation.

- **climatic analogues**
areas that are alike with respect to some of the major weather characteristics affecting the production of crops.

- **climatic factors**
light, atmospheric temperature, pressure and humidity, winds and other features of climate—regional, local and seasonal that influence vegetation.

- **climatic geomorphology (climatomorphology)**
the branch of geomorphology that concentrates on the influence of climate in the development of landforms. Climate is seen as directly influencing landforms as it determines the processes that operate, for e.g., the type, nature, and extent of weathering.

- **climatic province**
an area characterised by a general similarity of climate throughout.

- **climatic region**
an area on the earth's surface over which the combined effect of climatic factors result in an approximately homogenous set of climatic conditions. There have been many different divisions of the Earth into climatic regions using different climatic criteria but the four basic climatic regions are recognised as tropical, subtropical, temperate, and polar.

- **climatic stratification**
adaptation of species to a particular climate.

- **climatic variation**
the gradual change in the climate of a given locality occurring over an extended period of time, usually measured in centuries.

- **climatic year**
a continuous twelve-month period during which occurs a

complete annual climatic cycle and it is meant for the presentation of data relative to meteorological phenomena.

- **climatogenic soil classification**
  a genetic soil classification by taking climate as a variable, and assuming all other factors of soil formation as constant.

- **climatogenic structure (soil)**
  is formed due to the volumetric changes in soil (compression and compaction) during drying and freezing of water. This structure is characterised by a clearly expressed geometrical shape of crumbs and well-formed faces and edges. It may be said to exist in every soil, at least in lower horizons. The stability of this structure is entirely dependent on the presence of organic substances in the horizons where it is formed.

- **climatograph**
  a circular graph for depicting seasonal temperature conditions. Monthly temperatures are plotted from a centre with the aid of a graduated table.

- **climatology**
  the science which treats the component elements of weather and climate, such as temperature and rainfall, of their distribution over the earth and of the factors which determine and control their distribution.

- **climatope**
  the climatic factors of a place constitute the climatope.

- **climax**
  a plant community of the most advanced type that is capable of development under the prevailing climatic and soil conditions.

- **climax biotic**
  a climax which differs from the climatic climax of an area owing to the action of biotic factors.

- **climax climatic**
  a climax which owes its distinctive characters to climatic factors in conjunction with only such biotic influences as plants and animals naturally occurring in the area.

- **climax edaphic**
  a community which differs from the climax of the area owing to the influence of special soil factors.

- **climograph**
  a diagram in which two climatic elements for e.g., temperature and precipitation of a particular place are plotted against each other on a graph

(one as abscissa and the other as ordinates). The shape and position of the graph so obtained, provides an index of the general climatic character of the place under consideration.

- **climosequence**
  the various soil properties discussed in relation to moisture and temperature regimes form a continuous pattern over the landscape.

- **cline**
  a graded series of characters exhibited by a species or other related group of organisms, usually along a line of environmental or geographical transition. The populations at each end of the cline may be substantially different from each other.

- **clisere**
  a succession of climax communities in a given area, each giving way to the next as a result of climatic changes.

- **clod**
  a compact, coherent mass of soil ranging in size from 5 to 10 millimetres to as much as 200 to 250 millimetres, produced artificially, usually by the activity of man in ploughing, digging, etc., especially when these operations are performed on soils that are either too wet or too dry for normal tillage operations.

- **coupling agent**
  a solvent that has the ability to solubilise or to increase the amount of solubility of one material in another is referred to as a coupling agent.

- **covariance**
  the mean of the products of corresponding deviation of two variants from their individual means. Estimated as the ratio of the sum of cross-products, to the corresponding number of degrees of freedom.

- **cover**
  vegetative ground cover supplied by the above-ground portions of plants.

- **cover crop**
  a close-growing crop grown primarily for the purpose of protecting and improving soil between periods of regular crop production or between trees and vines in orchards and vineyards. Cover crops are known to reduce the danger of soil erosion and to protect the land from leaching losses.

- **cover ground**
  any vegetation producing a protecting mat on or just

above the soil surface. In forestry, low-growing shrubs and herbaceous plants under the trees.

- **cradle knoll**
  the earth raised and left in a knoll by an uprooted tree.

- **craze planes**
  are planes which are intermediate between nonpolar voids and skew planes. They are usually wider than skew planes and less congruent. This indicates that the adjacent peds do not interlock in good fit. They are commonest in soils of rather heterogeneous textural composition.

- **credit theory of the trade cycle**
  this theory explains the trade cycle in terms of changes in the volume of bank credit. It is said that there is a tendency for banks periodically to over-expand credit due to inter-bank competition. Each bank vies with the other to give advances to the customers. After a time, it is pointed out, the Banks become alarmed and so apply the reverse gear and start contracting credit. This undue expansion and contraction pf credit leads to vicissitudes in business activity. Against this theory, however, it is pointed *out* that changes in bank credit follow and do not precede changes in the level of business activity.

- **creep**
  the slow gradual downslope mass movement of soil and soil material slowly down steep slopes primarily by gravity, but facilitated by saturation with water and alternate freezing and thawing.

- **creep erosion**
  movement of soil to down steep slopes due to gravity.

- **creeping**
  of a stem, growing along the surface of the ground and rooting at the nodes.

- **creeping socialism**
  it refers to the gradual encroachment of the government on the private sector of the national economy.

- **creosote bush *(covillea tridentata)***
  an indicator plant that has good stand on nonsaline soils of sufficient depth to support specialised agriculture where water is available, poor stand usually indicate shallow soils underlain by rocks or hardpan.

- **cressa *(cressa truxillensis)***
  a plant, which is good indica-

tor of saline soils. More reliable than alkali heath. Require drainage and leaching.

- **critical concentration**
the limiting concentration of an impurity in an irrigation water after which it adversely affects crop growth, impairs the quality of the crop products, or damages the soil on which crops are grown.

- **critical constant**
a characteristic temperature, pressure and , specific volume of a gas above which it cannot be liquefied.

- **critical depth**
the depth of liquid flowing at the critical velocity in an open channel.

- **critical distance**
the distance from the crest to the point at which running water attains erosion velocity. It varies according to lithology, vegetation, climate, etc.

- **critical gradient**
hydraulic gradient at which the intergranular pressure in a mass of cohesionless soil is reduced to zero by the upward flow of water.

- **critical height**
the maximum height at which a vertical or sloped bank of soil will stand unsupported under a given set of conditions.

- **critical moisture**
the moisture content of either confined or unconfined soil samples attained by increase in the moisture content at a consistent rate because of deformations under constant load.

- **critical nutrient concentration**
the minimum concentration of a nutrient present in plant tissue at maximal growth.

- **critical percentage**
the point of supply of a given element at which no increase in yield will be obtained by an increase of the supply of that element to the plant. A decrease of the element will result in a decrease in yield.

- **critical period**
1. the shortest time span in the ontogeny of plant growth when a treatment will result in maximum effect.
2.period when the crop yield is affected due to inadequate moisture.

- **critical pressure**
the pressure of the liquid-vapour at the critical point.

- **critical properties**
physical and thermodynamic properties of materials at con-

ditions of critical temperature, pressure and volume, that is, at the critical point.

▪ **critical reaction**
in biology that pH at which a biological process becomes too slow to measure or at which organisms die.

▪ **critical region**
a test of a statistical hypothesis is made on the basis of a division of the sample space into two mutually exclusive regions. If the point falls into one (the region of acceptance), the hypothesis is accepted, if in other region (the region of rejection), it is rejected. Both regions are, in a sense critical, but it is customary to denote the second by the phrase critical region.

▪ **crop**
refers to plants sown and harvested by man.

▪ **crop acre**
is one used for producing some farm crop other than pasture.

▪ **crop associated weed**
non-parasitic weed associated with specific crop.

▪ **crop forest**
the entire collection of trees (including bamboos) growing on a given area. Forest crops may be classified in terms of their stages of development as seedling crops, thickets or sapling crops, pole crops, and tree crops.

▪ **crop growth rate (CGR)**
dry weight accumulated per unit of land area per unit of time.

▪ **crop insurance**
one of the means of getting protection against crop failure due to unforseen factors like bad weather, floods, fire, cyclones and hailstorm, etc.

▪ **crop intensification**
the increase of crop production per unit area, per unit of water, per unit of radiation, *etc.*, per unit time for a given geographical region.

▪ **crop irrigation ratio**
the ratio of irrigated area under a crop to the-total area under the same crop.

▪ **crop logging**
is a graphic record of the progress of the crop. It contains a series of physical and chemical measurements indicating the general condition of the plants to suggest changes in farm management for maximising crop yields. For crop logging sugarcane plants

are sampled every 35-days and analysed.

- **crop main**
  the stand left after thinning.
- **crop oil**
  petroleum oil with 14-25% aromatic content, used for selective weed control.
- **crop production strategy**
  a subset of the farm production strategy that involves one crop as its components. It is described by the spatial and temporal arrangement of crops to be raised in a parcel of land, and the procedure of growing each component crop (compartment technology).
- **crop requirements**
  a term sometimes used in considering agricultural water uses and refers both to net irrigation requirements.
- **crop residue**
  the portion of a plant or crop left in the field after harvest, or that part of the crop that is not used domestically or sold commercially.
- **crop residue management**
  use of crop residue for soil improvement or moisture retention.
- **crop residue recycling technology**
  the technology involves the incorporation of powdered groundnut shells or paddy husk at the rate of 50,00 Kg per hectare, one month in advance of sowing of sorghum, maize, pear] millet and groundnut crops. It has been developed for the management of red sandy loam 'Chalkas' soils which become very hard on drying.
- **crop rotation**
  is an order in which the chosen cultivated crops follow one another in a set cycle on the same field over a definite period for growth. The period may vary from one year to three years or even more. The rotations are planned carefully after considering the soil, climate, irrigation, the nature of demand in the market for agricultural produce and prices.
- **crop tolerance**
  the degree or the ability of the crop to be treated with a chemical but not injured.
- **crop tree**
  an individual tree which, because of species, form, and conditions, is selected for and is not likely to be present in the final harvestable crop, usually for sawlog purposes.

- **crop weather calendar**
a pictorial form, which provides information on the relationship between crops and weather conditions, they face at successive stages of plant growth.

- **crop year**
period from planting to harvesting of any crop.

- **crop yield**
quantity of harvested produce per unit area.

- **crop yield index**
refers to the overall productivity level attained on the farm in relation to the productivity level obtained in the area (locality). If the crop yield index of the farm under consideration is more than 100, it is said to be relatively more efficient than the average farm.

- **cropping intensity (cropping index)**
a method of expressing the number of crops grown on the same piece of land, for e.g., a monocrop is given a value of 100, a double crop of 200, and inter-cropping between rows in a full stand of a base crop, of 150.

- **cropping pattern**
yearly sequence and spatial arrangement of crops, or of crops and fallow, on a given area.

- **cropping scheme**
a plan of acreage of crops chosen for cultivation on a given farm area during an agricultural year.

- **cropping system**
cropping patterns used on a farm and their interactions with farm resources, other enterprises and available technology.

- **croppy**
term applied to jute fibre in which the top ends are comparatively hard and stiff.

- **cross cultivation**
the working of land by ploughing second time at right angles, to the first.

- **cross drill**
a type of sowing for uniform distribution in which half the quantity of seed is drilled in one direction and the other half drilled at right angles to first.

- **cross elasticity of demand**
when two goods, which are close substitutes for each other, are in competitive demand, an increase in the price of one will increase the de-

mand for the other. The effect of this change in the price of one good on the quantity demanded of the other is referred to as cross elasticity of demand.

- **cross fertilisation**
  union of an egg cell from one individual with a sperm cell from another.

- **cross inoculation group**
  are leguminous plants of one or more genera or species which develop root nodules in association with the same varieties or species of Rhizobium.

- **cross protection**
  the phenomenon in which plant tissues infected with one strain of a virus are protected from infection by other strains of the same virus.

- **crossing over**
  the exchange of material between homologous chromatids by the formation of chiasmata.

- **cross-section**
  the apparent cross-sectional area of a nucleus as calculated on the basis of the probability of occurrence of a reaction by collision with particles. It may or may not coincide with value for the geometrical cross-section area - $r^2$.

- **crotonylidene diurea**
  the acid catalysis of either urea acetaldehyde or urea-crotonaldehyde condensations yields a ring structured compound, CDU, containing two ring and two side-chain nitrogen atoms. A fertiliser source of slowly available nitrogen. It contains 28% N of which about 10% is present as $NO_3$-N.

- **crotovinas**
  channels in the soil, caused by burrowing animals of all types, and which have become filled with surface soil and/or organic matter.

- **croute calcaire**
  croute calcaire, or hardened caliche, is often found in thick masses by only a few inches of soil. The common Cca-horizons of Chernozems are easily conceived as part of the soil profile, although they are not within the solum.

- **crow bar**
  a tool used for digging and turning clods.

- **crucifera boron deficiency**
  disease of crucifers caused by

deficiency of boron in soil. Characterised by brown necrosis in fleshy pith of stem and core of plants, internal breakdown of brown nectotic tissues. Control by applying 10 Kg. borax per acre.

- **crushing strength**
the force required for crushing a mass of dry soil or, conversely, the resistance of the dry soil mass to crushing. Expressed in units of force per unit area.

- **crust**
a surface layer on soils that is much more dense and more compact, hard, and brittle.

- **cryic**
one of the soil temperature categories. Mean annual soil temperature 0 to 8°C with summer temperatures less than 15°C (a number of specific requirements).

- **cryology**
the study of snow, ice, hail, sleet—in fact, of any water at low temperature.

- **cryophytes**
plants growing on ice and snow, micro-plants, largely consisting of algae but including also some mosses, fungi and bacteria. Algal forms may be so abundant as to colour substratum, for e.g., 'red snow' due to presence of *Chlarny-domonas.*

- **cryptomerism**
a failure of characters to show in the offspring, which nevertheless contain the corresponding hereditary factor.

- **cryptophyte**
a plant perennating by means of rhizomes, corms, or bulbs underground, or by under water buds.

- **cryptoplasm**
that part of the cytoplasm that appears to lack granules.

- **cryptopodsol**
a soil, which does not appear to be a podsolibut, can be shown to be so by chemical analysis.

- **crystal axis**
a reference axis used for the vectoral properties of a crystal.

- **crystalline materials**
  materials with a regular microstructure. Full crystallinity implies regularity in three dimensions, clay minerals are usually of limited two-dimensional crystallinity.

- **crystalpedoturbation**
  soil mixing by growth of crystals, such as of halite (NaCl).

- **CS**
  a subhorizon symbol indicating an accumulation of calcium sulphate. Calcium sulphate accumulations commonly occur in the C below ca accumulations in chernozemic soils but may occur in other horizons as well. Before the symbol cs is used, the horizon must have more sulphates than the parent material is presumed to have had.

- **cubic foot per second**
  a unit of discharge for measurement of flowing liquid, equal to a flow of 1 cum ft per sec past a given section.

- **cubic metre per second**
  a flow of water equivalent to a stream 1 metre wide and 1 metre deep flowing at a velocity of 1 metre per second.

- **cull**
  an inferior plant rejected from nursery stock trees or logs that are of merchantable size but are rendered unmerchantable by defects.

- **culm**
  the jointed stem of a grass which is usually hollow except at the nodes or joints.

- **cultivar**
  any agricultural or horticultural 'variety'. The term is derived from the words cultivated variety.

- **cultivate**
  1. to till the soil.
  2. to loosen the soil and remove weeds among growing plants.
  3. to plant, tend, harvest and improve by plant breeding.

- **cultivation**
  1. a mechanical stirring of the soil in place as in seed bed preparation or weed control.
  2. Cultivation generally helps improve aeration and infiltration of water thus enhancing plant growth.

- **cultivation index**

the effect of cultivation and cultivation practices on the soil organic matter level is expressed as cultivation index which amounts to organic carbon content of cultivated soil expressed as a percentage of the organic carbon content of the virgin soil. Such an index is designated Cx for organic carbon and nx for nitrogen. In majority of cases it lies between 50 to 70 irrespective of the period of cultivation—suggestive of the balance state attained.

- **cultivation sheet**

a record which provides the complete history of the individual crop grown on the field with its cost of cultivation from preparing the land to harvesting the crop and all other extra charges up to sale, showing at the end total expenses of the crop, yield, and net profit or net loss.

- **cultivation shifting**

a method of cyclical cultivation, chiefly in vogue in the tropics, where cultivators cut the tree crop, burn it, and raise field crops for one or more years before moving on to another site and repeating the process. Soil fertility is restored and grassy weeds are eradicated.

- **cultivator**

1. an implement consisting of a number of tynes attached to a frame which is used for tilling the soil between standing rows of crops. It stirs the soil and breaks the clods. The tynes fitted on the frame of the cultivator enter the soil to the required depth and comb the soil deeply in the field.

2. A cultivator performs functions intermediate functions between those of plough and the harrow. Destruction of weeds is the primary function of a cultivator.

- **cultivator (tyne type)**

c shaped spring steel shanks are mounted in rows, usually about 30 cm apart. Chisels or sweep shovels are mounted on these shanks. Heavy duty types are made for deep chiselling but lighter types with sweeps can be used for seed bed preparation and weeding.

- **cultivator (wide blade)**
'V' shaped sweeps 90 to 240 mm wide with the sweeps angle from 72 to 100 degrees.

- **culturable waste**
1. class of land under revenue classification, infertile land, which has either been not taken up for cultivation or abandoned after a few years of cultivation, when it becomes uneconomical
2. sub-marginal land which cannot be brought under cultivation or cultivated in an economical manner.

- **cultural**
1. pertaining to, or produced by, plant breeding.
2. pertaining to tillage practices.

- **cultural control**
control measures including modifications of the planting, growing, cultivating, and harvesting of crops aimed at prevention of pest damage rather than destruction of an existing infestation.

- **cultural operations (forestry)**
a general term for operations, as a rule not directly remunerative, undertaken to assist or complete existing regeneration to promote the proper development of the tree crops or to minimise the after-effects of felling damage.

- **cultural practices**
practices adopted in tillage, harvesting and improvement of crop plants.

- **culture**
the growth of organisms, especially microorganisms, under clearly defined conditions which are usually artificial. Such methods may be used experimentally, for e.g., with fungi and bacteria.

- **cumulative effect**
1. increase in the ultimate fertility of lands of certain types of soils under certain environmental conditions because of repeated applications of organic manures.
2. the result of some poisons which build up or are stored in the body so that small amounts eaten or contacted over a period of time can sicken or kill an animal or person for e.g., mercury compounds, thallium sulphate etc.

- **cumulative frequency curve**
a frequency curve in which values are added successively to each other and then converted into percentage.

- **cumulative pesticides**
those chemicals which tend to accumulate or build up in the tissue of animals or in the environment (soil, water).

- **cumulative runoff**
the total summation of runoff over a specified period of time. Successive summations are frequently plotted against time.

- **cumulisation (fr. *cumulus,to* heap)**
the accumulation of mineral material onto the surface of the soil by either air or water. The effects of this process are most evident in digressional areas where material eroded from soils upslope has accumulated.

- **cup conductance (tensiometer)**
the rate with which water is transferred from soil to cup and cup to soil, depending upon the conditions. It is desirable to have high cup conductance to have short response time.

- **cure**
to prepare for preservation by drying or other processes.

- **curing kiln or barn (tabacco)**
used for curing tobacco, usually operate on the principle of applying heat at the bottom and have top and bottom air vents. They may be built from logs, sawn timber, galvanised iron, asbestos, cement, brick or reinforced concrete. The heating system may operate on wood burners or liquid fuel burners. Forced circulation down draught units are also common.

- **current fallows**
cultivable land which is not cultivated during the current year.

- **curvature**
a change in the direction of growth of part of a plant, due to one side growing faster than the other.

- **curve of demand**
a graph showing the amount of water needed for irrigation at various tirfies daring the crop season based on elements of time and quantity.

- **cusec**
is the quantity of water flowing at the rate of 1 cu. ft. per second. A cusec flowing for one hour is approximately one acre inch of water. It takes

about 22.4 hours to deliver one acre inch of water at a discharge rate of 1000 gallons per hour. one cusec equals 538,000 gallons per day.

- **cush cush**
  used in sugar manufacture for fine begasse screened from raw juice.

- **cut back**
  1. to prune a plant by removing the terminal ends of shoots or branches.
  2. to reduce to the original number, size, etc.

- **cut stump treatment**
  application of a brush killing chemical to freshly cut stump of unwanted tree.

- **cutans**
  a modification of soil structure, texture or fabric of natural surfaces in soil materials due to concentration of particular soil constituents or in situ modification of the soil plasma, cutans can be composed of any of the component substances of the soil material.

- **cuticle**
  the outer corky or waxy covering of the plant formed by oxidation of plant oils.

- **cuticular diffusion**
  the passage of gases through the cuticle of a plant.

- **cuticular transpiration**
  the loss of water-vapour from a plant through the cuticle.

- **cutie pie**
  a 'Gun-type', portable instrument for measuring the radiation level.

- **cutin**
  the substance of plant cuticle deposited on, or in, the outer layer of cell walls. It is formed by the oxidation and condensation of fatty acids.

- **cutinisation**
  the formation of cutin.

- **cutoff ratio**
  the ratio of the length of a cutoff to the original length of the river.

- **cytophagy**
  the engulfing of cells by other cells.

- **cytoplasm**
  all the protoplasm of a cell excluding the nucleus. It is usually a transparent gel or viscos fluid with inclusions of various sizes ranging from microscopically visible plastids and Golgi apparatus to invisible ribosomes.

**cytoplasmic membrane**
the inner-most covering of the bacterial cell. It surrounds the cytoplasm and shrinks with this material during plasmolysis. It reacts to tests for lipids and proteins. It always initiates division of the cell and is semi-permeable.

**cytosine**
a pyrimidic base, $C_4H_5N_3O$. Exists in the fraction of the humus, which is not precipitated by acids.

**cytosol**
the soluble fraction of cytoplasm remaining after all particles have been removed by centrifugation.

**cytotaxonomy**
the use of chromosome number, size, and shape in the classification of organisms.

**d trials**
a series of large plot trials for comparison of varieties using farm methods of seeding and harvesting.

**daily mean soil temperature**
the average of the values observed during a day (at any regular intervals or maximum and minimum) of soil temperature.

**dal soil**
a local name for a peat soil in the Netherlands, consisting of a sandy subsoil covered with a layer of lumps of young peat.

**damaged seedlings**
seedlings without cotyledons, seedlings with splits, cracks or lesions, and seedlings without a primary root.

**damp soil**
soil which is apt to be too damp.

**damping off**
a disease which attacks young seedlings at ground level causing them to rot and fall over. It is usually caused by species of the fungus pythium, and is favoured by overcrowding in damp conditions.

**danielli membrane**
theoretical structure of the cell membrane put forward by Danielli, consisting of a double layer of phospholipid, with protein arranged along each outer edge.

**darcy weisbach formula**
a formula for determining the loss of head in flowing water. The loss of head is equal to a coefficient that varies with the surface roughness of the conduit and the Reynolds' number, the length of the conduit divided by its diameter times

the velocity head of the flowing fluid. In the case of a noncircular conduit or a circular conduit not flowing full, four times the hydraulic radius as substituted for the diameter.

- **dark field microscopy**
1. a type of microscopic examination in which the microscopic field is dark and any objects such as microorganisms are brightly illuminated. 2. In microbiology it is usually employed for detection and identification of living and dead microorganisms, enumerating microorganisms and studying their structure and morphology.

- **dark reactions**
a group of reactions that follow the light reaction in photosynthesis and form glucose and other reduced products from carbon dioxide. They are not dependent on light, although they can take place in the light.

- **dark seed**
a seed which will germinate, only if kept in the dark, when other conditions would normally favour germination.

- **datum**
a set of parameters and control points used to accurately define the three dimensional shape of the Earth. The corresponding datum is the basis for a planar coordinate system.

- **davies compactometer**
a needle with a spring and graduations making it possible to measure the resistance of a soil to penetration, and therefore its compactness.

- **day length**
the number of hours of light for each 24-hour cycle. It is an important factor in the growth of plants.

- **day neutral plants**
plants which flower after a period of vegetative growth, regardless of the photoperiod, for e.g., *Lyco-persicum esculentum* (tomato), peas *(pisum sativum)*, etc.

- **day position**
the situation adopted by leaves of plants during the day, they alter this position when the light intensity decreases.

- **day sleep**
the folding of leaflets of a compound leaf in bright light. This brings the stomatal surfaces together, thus probably reducing the water-loss.

- **days to harvest**
the minimum number of days permitted by law between the

final pesticide application and the harvest of crops.

▪ **days to slaughter**
the minimum number of days established by law, between the final pesticide application and the date an animal is to be slaughtered.

▪ **DE**
a prefix implying removal.

▪ **dead soil**
soils having top layer of peat lump.

▪ **dead time**
the length of time immediately following an impulse that the instrument remains insensitive and unable to record another ionising event.

▪ **dead water**
the fraction of the water in the soil which is so strongly held that it is unusable by plants.

▪ **dealkalisation**
the removal of the na⁺ ions from the exchange sites usually by treatment and leaching.

▪ **death point**
the maximum, or minimum limit of any factor in the interval or external environment, beyond which the cell organism is killed.

▪ **debasification**
the removal of the bases from the absorption complex, a consequence of leaching by slightly acidified waters.

▪ **debpe**
a unit of electric dipole moment equal to 3.33564 x $10^{13}$ coloumb meter. It is used in expressing the dipole moment of molecules.

▪ **debris**
1. an organic waste from dead or damaged tissue.
2. the coarse material resulting from the decay and disintegration of rocks.

▪ **debye falkenhagen effect**
the increase in the conductance of an electrolytic solution when the applied voltage has a very high frequency.

▪ **deca**
a Greek prefix meaning 'ten'. For example 1 decametres (dem) = 10 metres (m).

▪ **decade sealer**
a recording instrument hav-

ing a scaling factor of ten.

■ **decalcification**

1. loss or removal of calcium or calcium compounds from a calcified material such as soil by leaching.

2. Technically, replacement of monovalent metallic ions such as $Na^+$ by H+ ions.

■ **decarbonation**

the decomposition of calcium carbonate under the action of precipitation water which contains dissolved $CO_2$ and humus. The calcium is carried off or becomes bound to the humus.

■ **decay**

1. the destruction of organic materials through the action of enzymes produced by microorganisms.

2. the radioactive disintegration of a molecule.

■ **decay constant**

synonym for disintegration constant.

■ **decay series**

synonym for decay chain. A series of radioactive decompositions in which radioactive daughters are produced, these radioactive daughters then becoming parents by producing radioactive daughters of their own.

■ **decayed seedling**

seedlings with any of the essential structures so diseased or decayed that normal development is prevented, except when there is clear evidence that the cause of infection is not the seed itself.

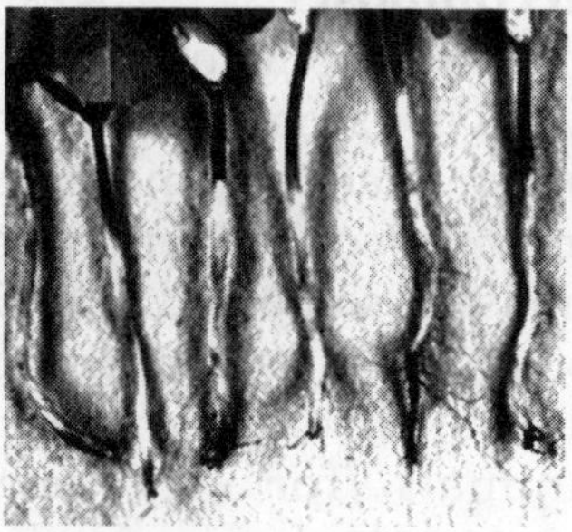

■ **deciduous**

plants or trees that shed leaves or wans at a particular season or stage of growth.

■ **decimation**

destruction of a considerable part.

■ **decinormal**

a chemical solution that has one-tenth normality in reference to a 1 normal solution.

■ **declivity**

a downward sloping land surface.

■ **decoction**

something obtained by boiling a substance to remove its

soluble constituents.

▪ **decolourising carbon**
porous or finely divided carbon (activated or bone) with large surface area, used to adsorb coloured impurities from liquids.

▪ **decomposer**
a heterotrophic organism in an ecosystem that breaks down dead organic matter into its constituent parts, utilising some and releasing others to be used by the producers. Decomposers include animals (for e.g., earthworms), bacteria, and plants (fungi).

▪ **decomposition**
the more or less permanent breakdown of a molecule into simpler molecules or atoms.

▪ **decomposition potential**
the electrode potential at which the electrolysis current begins to increase appreciably.

▪ **decontaminate**
to make safe by removing any pesticide from equipment or other surfaces as directed on a pesticide label or by an agricultural authority.

▪ **deep litter system**
system of keeping poultry in a house on the floor of which is placed litter composed of wood shavings, sawdust, straw, etc. mixed with hydrated lime to about 6 inch depth, dropping not removed, although litter is stirred periodically and fresh material added. Litter is replaced after a year, sometimes 2 or 3 years.

▪ **deep percolation**
1. the moisture which penetrates below the depths from which it may not be used by plants, it represents that part of the water absorbed which exceeds the field capacity of the soil within the depth of root development.
2. in a geologic sense, the amount of water that is lost from a basin by leakage through the geologic formation.

▪ **deep percolation loss**
water that percolates downward through the soil beyond the reach of plant roots.

▪ **deep placement**
is a method of applying nitrogenous fertilisers in which the fertilisers are placed directly in the reducing zone of the paddy soil.

▪ **deep seepage**
that portion of the runoff which escapes from a reser-

voir through the underlying earth or rock strata, below any possible intercepting cut-off constructed at the dam.

- **deep soils**
soils characterised by a root zone for fibrous roots in excess of 75 cm.

- **deep water rice**
also called floating rice. special floating variety of rice grown in low-lying areas in 15 to 20 feet of water. Sown broadcast with the pre-monsoon rains, grows as the level of water in the field rises gradually, keeping up with its rising level.

- **deep wells**
a well drilled or bored in a deep ground water source. A galvanised iron tube or pipe is driven into the bore by a suitable device. These wells may vary in depth from 85 to 100 feet or more. Bore or tube wells come under this category.

- **deepening**
expresses a decrease in the atmospheric pressure at the centre of low pressure system.

- **defective per cent**
the number of defective units in a lot divided by the total number of units times 100.

- **defective unit**
in quality control, a unit which does not reach some prescribed standard and is therefore rejected.

- **deferred grazing**
withholding animals from a pasture beyond the normal beginning of the grazing season.

- **deferred shoots**
shoots arising from dormant buds.

- **deferrisation**
loss of iron by migration, affecting the entire soil, or one horizon, or a part of a horizon. probably a result of leaching by organic or tannic acid solutions, and is followed by an at least temporary conversion of iron to the ferrous state.

- **deficiency**
a type of mutation caused by loss of one or more nucleotides from a DNA segment. Deficiencies can be very large, encompassing many genes and megabases of DNA, to the point of producing a visible cytological abnormality in a chromosome.

- **deficiency diseases**
pathological condition in plants and animals, due to lack

of certain necessary nutritive substances, for e.g., crown rot in sugar beet due to boron deficiency.

- **deficit financing**
  it is a practice resorted to by modern governments of spending more money than they receive in revenue. It means deliberately budgeting for a deficit. The deficit budget is incurred by the government either to deal with a depression and serious unemployment as in western capitalist countries or to break the vicious circle of poverty and under-development as in underdeveloped countries. This practice is highly controversial.

- **definite composition law**
  1. the law that a given chemical compound always contains the same elements in the same fixed proportions by weight.
  2. Also known as definite proportion law.

- **definition**
  the definition identifies the essence of a concept and is comprised of the genus and differentia.

- **deflation**
  1. removal of particles or fine gravel by the wind. It is exercised only on soils entirely or partially devoid of plant cover, either naturally (arid or cold deserts) or artificially (crops, clearing). It is one of the principal factors of soil erosion.
  2. a fall in the general price-level associated with a contraction of the supply of money and credit.

- **deflationary gap**
  a certain volume of investment, public and private, is necessary for an economy to attain full employment. If the economy does not attain full employment, then the deflationary gap is the amount of additional investment required to restore employment to the level of full employment.

- **deflocculate**
  1. to separate the individual components of compound particles by chemical and/or physical means.
  2. to cause the particles of the dispersed phase of a colloidal system to become suspended in the dispersion medium
  3. to separate or break down soil aggregates into their component particles.

- **defluorinated phosphate**
is produced by the treatment of mineral phosphates with silica and water vapour in furnaces and kilns, whereby most of the fluorine is driven off and the phosphorus is converted into forms that are soluble in neutral ammonium citrate solution, but practically insoluble in water.

- **defoliant**
any substance or mixture of substances for which the primary use is to cause the leaves or foliage to drop from a plant.

- **defoliation**
a reduction in the normal amount of foliage due to insect or fungal attack or other injury or by spraying, may be partial or complete.

- **defoliator**
a compound which causes the leaves or foliage to drop from the plant.

- **deforest to**
to remove the tree crop from a piece of land without the intention of reforesting.

- **deformation**
alteration of shape.

- **deformation elastic**
deformation such that the deforming force ceases to act, the object resumes its initial form. The elastic deformation of soils is proportional to the pressure exerted, it is often accompanied by an elimination of water.

- **deformation plastic**
deformation such that when the force ceases to act, the object remains deformed.

- **deformed seedlings**
seedlings with weak or unbalanced development of the essential structures such as swollen roots and stunted roots, split plumules or coleoptiles without a green leaf.

- **degenerated**
that which has lost its ancient or ancestral characters.

- **degenerated soil**
a soil which, after having reached maturity in a certain line of development, is subjected to new conditions of plant cover, drainage or climate, and losses all or some of its characters.

- **degleification**
opposite to gleyification. The possible reoxidation of ferrous salts, and especially reconstitution of porous structure, resumption of microbial activity ensuring the fertility of the soil. It may result from

land improvement, from cultural steps and from natural vegetation. It is very slow and gradual.

- **degradation**
chemical or biological breakdown of a complex compound into simpler compounds.

- **degraded alkali soil**
as used by de Sigmond, this term refers to a soil containing less than 0.1 per cent salt but with appreciable quantities of exchangeable sodium and hydrogen. Soil pH values in the surface horizons may be below 6.0 but are usually between 6.5 and 8.0. This corresponds approximately to the great soil group known as soboth.

- **degraded chernozem**
a zonal great soil group consisting of soils with a very dark brown or black $A_t$ horizon underlain by a dark gray, weakly expressed $A_2$ horizon and a brown B horizon, formed in the forest-prairie transition of cool climates (1949 system).

- **degraded ecosystem**
an ecosystem of fluctuating communities resulting from recent major disturbance and containing few trophic levels. It has a less complex food web than a neutral ecosystem.

- **degraded soil**
a soil that has been subjected to stronger leaching than the type soils of the area, hence lower in fertility.

- **degree brix**
measure for sugar content of sugarcane or sugar solution measured by a special instrument, saccharimeter.

- **degree days**
the amount of heat energy that has been absorbed by the soil over a given period of time.

- **degree of aggregation**
measure of the proportion in which aggregates are present.

- **degree of cation saturation of the soil (v)**
it is the ratio V= 100 S/T of the sum S of the content of exchangeable cations other than H, to the total exchange capacity T, both evaluated in milliequivalents, V=90=100% for saturated soil.

- **degree of exposure**
the amount or extent to which a person has been in contact with a toxic pesticide.

- **degree of freedom**
a comparison of items within a body of data, independent of any other comparison used in

the analysis. The number of degrees of freedom is the number of independent comparisons that can be made within the body of data.

- **degree of humification**
the ratio of the humic acids to the total carbon.

- **dehulled seed**
seed from which the pods, glumes or other outer covering have been removed, as sometimes with lespedeza and timothy, also often ambiguously referred to as 'hulled' seed.

- **dehusk**
to separate husks from seeds, grains, etc.

- **dehydrase**
an enzyme that catalyses the removal of water, *i.e.* condensation.
For e.g., $2C_6H_{12}0_6=C_{12}H_{22}0_u + H_20$.

- **dehydration**
the rapid removal of moisture to a very low level.

- **dehydration curve**
a curve obtained by plotting temperatures on the abscissa and the per cent of moisture lost on the ordinate.

- **dehydrohumic**
an alkali soluble component of humic substances.

- **delayed action**
with some herbicidal chemicals delayed response is expected. Considerable time may elapse before maximum effects can be observed. Usually treated plants stop developing soon after treatment, then gradually die.

- **deletions**
loss of a section of the genetic material from a chromosome. The size of the deleted material can vary from a single nucleotide to sections containing a number of genes.

- **delinting**
removal of the lint fibres that cling to cotton seed after ginning. Can be done mechanically on special gins or by treatment with sulphuric acid. Makes disease control easier, hastens germination and permits more accurate spacing of seed when planting.

- **deliquescence**
the process that occurs when the vapour pressure of the saturated aqueous solution of a substance is less than the vapour pressure of the water in the ambient air. Water vapour is collected until the substance is dissolved and in equilibrium with its environment

- **deluge**
  a sudden flooding or inundation of water.
- **deluvial**
  carrying off earths towards the low parts.
- **demand function**
  in a given market, in a given period, the demand function for a commodity is the relation between the various quantities of the commodity that might be bought and the determinants of those quantities.
- **demand schedule**
  is a list of the various quantities of a commodity which an individual consumer purchases at different alternative prices in the market.
- **demersus**
  Latin word meaning 'subaquatic'.
- **demography**
  the statistical study of human populations, especially with reference to size and density, distribution and vital statistics.
- **demonstration**
  the procedure of doing something in the presence of others either as a means of showing them how to do it themselves or in order to illustrate a principle.
- **demonstration effect**
  it refers to the impact produced on an individual consumer when he sees a large variety of superior goods and decides to increase his expenditure on these goods even though his income has not changed (or, increased). According to James S. Duesenberry, the demonstration effect amends the Keynesian theory of consumption function, according to which an individual's consumption expenditure is related only to the level of his income.
- **demonstration, result**
  a project carried on for a substantial period of time under the supervision of a professional worker to show the value of a recommended practice, during which time records are kept and comparisons made.
- **denaturation**
  the process of partial or total alteration of the native structure of a macromolecule resulting from the loss of tertiary and secondary structure that is a consequence of the disruption of stabilising weak

bonds. Denaturation can occur when proteins and nucleic acids are subjected to elevated temperature or to extremes of pH, or to non-physiological concentrations of salt, organic solvents, urea or other chemical agents.

- **dendrograph**
  an instrument which is used to measure the periodic shrinking and swelling of tree trunks.

- **dendron**
  Greek word meaning 'tree'.

- **denitrifying bacteria**
  a group of soil bacteria which break down nitrites and nitrates anaerobically to produce nitrogen.

- **denizen**
  a species of plant which can maintain itself in the wild, but was probably introduced by man.

- **dens**
  the Latin for 'tooth'.

- **dense**
  1. the close packing of the component parts in a substance.
  2. designating a soil mass in which pore spaces form a relatively small proportion to the total volume and where any large pores or cracks are absent.

- **density**
  the ratio of the mass of a substance to the volume it occupies. In oceanography, it is equivalent to specific gravity and represents the ratio of the weight of a given volume of sea water to that of an equal volume of distilled water at 4.0Â°C or 39.2Â°F.

- **density dependent**
  having influence on individuals in a population that varies with the number of individuals per unit area in the population.

- **density gradient centrifugation**
  separation of particles according to density by employing a gradient of varying densities, at equilibrium each particle settles in the gradient at a point equal to its density.

- **density independent**
  having influence on individuals in a population that does not vary with the number of individuals per unit area in the population.

- **density of cover**
  the percentage of the ground surface covered by vegetation or by a given species.

- **density of population**
  number of individuals in a population relative to space or

specific factors affecting economic development.

- **denudation**
refers to the complete destruction of trees and humus in a forest. It results in quick run-off rain water which leaves the ground-water springs unregenerated.

- **denuded quadrate**
a square piece of ground, from which all the vegetation has been removed to study its decolonisation.

- **deodorant**
a substance which removes and neutralises offensive odours, generally by oxidising the products of decomposition of organic matter and thereby absorbing or masking the odours.

- **departure**
the amount (positive or negative) by which the value of a meteorological element departs from normal average or expected value for a given place and a time. It is also termed as anomoly.

- **depauperate**
plant which does not grow to its natural size and has a starved or underdeveloped appearance.

- **dependent plant**
a plant which does not make its own food but depends on other living or dead things for its nutrition.

- **dependent variable**
a variable whose magnitude depends on, or is a function of, the value of another variable (or other variables), a variable whose value is being estimated from that of one or more independent variables to which it is related, when represented graphically, the 'Y' or vertical axis is conventionally used for the dependent variable.

- **dephosphorated slag**
a phosphatic fertiliser made from iron ore. Limestone is added to molten iron ore rich in phosphorus and oxygen blown into mixture. The slag floating on the top is cooled and ground. It contains phosphorus in the form of tetracalcium phosphate ($Ca_4P2O_9$) and calcium silicophosphate ($Ca_5SiP2O_{12}$). Strength : 17-23% total $P_2O_5$ of which 15-20 % is soluble in 2% citric acid.

- **deplasmolysis**
the recovery from plasmoly-

sis by the intake of water by osmosis.

- **deplated soil**
any soil that has been so used or misused that the resultant loss of fertility has impaired its economic utility for crop production.

- **depletion**
general term for washing out or eluviating soluble materials from the solum.

- **depletion curve**
that part of a hydrograph extending from the point of termination of the recession curve to a subsequent rise or alternation of inflow, due to additional water becoming available for stream flow, that part of the hydrograph representing the rate of water flow or seepage from ground water storage into the stream channels.

- **deposit**
material left in a new position by a natural transporting agent such as water, wind, ice, or gravity, or by the activity of man.

- **deposit builder**
chemical that aids the toxicant to stick to the target surface.

- **deposition**
1. the accumulation of soil material dropped because of slackening movement of the transporting agent— water or wind. Accumulation at the foot of an eroded slope, alluvial fans and dunes are examples of deposition.
2. in pesticide application, the dust or spray which remains on the plant surface.

- **deposition phases**
phases for soil depositions are as follows:
a. overblown phase. The deposit of wind-removed material lying on the soil is great enough to influence management but not great enough to destroy the essential characteristics of the soil series. Such a soil takes its textural class name from that of the overlying material.
b. wind-hummocky phase. Recent wind deposits lie on the soil in a fine pattern of hummocks that markedly alter management requirements of the soil but do not obliterate, the essential characteristics of the soil series.
c. overwash phase. Deposits from water erosion lie thick enough on the soil to influence management requirements significantly but are not deep enough to destroy the essential characteristics

of the soil series. Such a soil takes its textural class name from that of the overlying materials. These may be named overwash phases, sanded phases, or rock-wash phases. Ordinarily they are not needed on very young alluvial soils.

- **depression**
it is a term used to describe a period of heavy unemployment and stagnation of business activity. It is characterised by a sharp decline in production, little capital investment, concentration of credit, falling prices, a very high rate of business failures and mass unemployment. The biggest depression in the world's history was that of 1929-1933.

- **depression storage**
the holding of water in depressions is called depression storage or surface water retention.

- **depth area curve**
a curve showing the relationship between the average intensity of the rainfall of a storm and the area of the storm.

- **depth factor**
a depth which would contain enough grain that if all the theoretical heat available for drying could be used, it would all dry to equilibrium in a period of time equal to the time of one-half response.

- **derived data**
records of observations and measurements of physical facts, occurrences, and conditions which have been developed from basic data by means of standard methods of computation and estimate.

- **derived demand**
demand in which a good or service is wanted not for its own sake but for its contribution to the production of other goods, for e.g., the demand for labour or raw materials. Thus, the demand for such goods is dependent on (or 'derived' from) the demand for the final product.

- **dermal toxicity**
the ability of a pesticide chemical to poison an animal or human by skin absorption.

- **dermato**
prefix meaning the 'skin'.

- **dermatogen**
the external layer of a stem or root apex, one cell thick, and

giving rise to the epidermis.

- **dermo**
prefix meaning the 'skin'.

- **desalination**
the process of removing salt from water. Most often, it is a distillation process in which salt water is evaporated and then condensed, leaving the salt as residue. A variety of processes whereby the salts are removed from seawater, resulting in water that can be used for human consumption, salt removal from sea or brackish water.

- **desalination factor**
the concentration of dissolved substances in the influent water of a desalination process, divided by their concentration in the desalted water.

- **desalination processes**
distillation, multi-stage flash distillation, and the less common membrane processes of electro-dialysis and reverse osmosis. Membrane processes are more expensive than distillation, but this situation may alter.

- **desalinisation**
the artificial removal of the Na salts from a soil by leaching in the presence of Ca.

- **desert**
a region so devoid of vegetation as to be incapable of supporting any considerable population. Four kinds of deserts, may be distinguished:
a) the polar ice and snow deserts, marked by perpetual snow cover and intense cold
b) the middle latitude deserts, in the basinlike interiors of the continents, such as the Gobi, characterised by scant rainfall and high summer temperatures
c) the tradewind deserts, notably the Sahara, the distinguishing features of which are negligible precipitation and large daily temperature range, and
d) coastal deserts, where there is a cold current on the western coast of a large land mass, such as occurs in Peru.

- **desert crust**
a hard layer, containing calcium carbonate, gypsum, or other binding materials exposed at the surface in desert regions.

- **desert pavement**
the layer of gravel or stones left on the land surface in

desert regions after the removal of the fine material by wind erosion.

- **desert saltbush *(atriplex polycarpd)***
an indicator plant. Where stands are pure with good growth, the soils are nonsaline or only slightly saline and are suitable for irrigation. Where growth is poor, there may be a hardpan or salt in the subsoil.

- **desert soils**
are zonal soils found in arid-tracts where even grasses can not persist. They are of little agricultural value except under irrigation, when they become highly productive. Generally, there is poor development of profile in such soils.

- **desert strip water harvesting system**
desert strip cropping uses water harvested from a collector area to help in supplying the moisture requirements of a cultivated crop on a smaller farmed area. An advantage of leaving the collector area in its natural state is that it can be used in its traditional manner for livestock raising. If the crop is a failure, only a small amount of land is lost to the livestock raising activity. If a crop is produced the collector area helps in providing the moisture for the crop as well as feed for the livestock.

- **desert varnish**
a glossy sheet or coating on stones and gravel in arid regions.

- **desertification**
the ecological changes that sap land of its ability to sustain agriculture and human habitation. Deserts are the product of a dry climate, and they prevail where global patterns of atmospheric circulation dictate that rain will be absent. Deforestation is the main cause of desert spread.

- **desi plough**
a local plough usually made of wood but often with a metal tip on the part which cultivates the soil either by trenching and/or partially turning the soil. Usually drawn by oxen.

- **desiccant**
a crop drying agent. For example, when seed alfalfa or red clover preharvest are sprayed with dinitro and oils, usually the foliage is killed by contact action, and often seed

moisture is reduced, which aids harvest.

- **desiccation**
  a complete or almost complete deprivation of moisture or of water not chemically combined.

- **desiccation resistance**
  the lowest relative humidity with which a plant can come to equilibrium without injury.

- **design**
  a method of arranging sample or experimental plots to minimise the effects of uncontrolled variations in fertility and other natural factors, and to make it possible to estimate the magnitude of such effects in relation to those due to variations in treatment *i.e.* to controlled variations.

- **desilting**
  periodical removal of silt (finer soil particles) which accumulates at the bed of a reservoir or a canal. Desilting is necessary to maintain the capacity of the canal or reservoir.

- **desludging**
  removal of sludge from a sedimentation tank, etc.

- **desorption**
  the phenomenon opposite to absorption. The release of the ions held on the colloids when the concentration in the surrounding solution is decreased.

- **dessalement**
  the removal of the sodium salts from soil which has been damaged by sea water. Liming and irrigation encourages restoration of a good structure.

- **destarched**
  when green plant no longer contain starch, they are called destarched. This is brought about by keeping the plant in the dark. This stops photosynthesis, and any starch present will either have been used, or translocated.

- **destructive distillation**
  the decomposition of organic compounds by heat in a closed container and the collection of the resultant volatile liquids and oils.

- **desuckering**
  removal of suckers

- **detailed soil map**
  a soil map in which the soil types and phases are mapped in the detail required to show all boundaries between map-

ping units, including areas of one unit within another, that are significant to potential use.

- **detassel**
to remove the tassel.

- **detasseling**
the removing of the tassels in maize plants meant to serve as female parent in hybrid seed production.

- **detergent**
a cleansing agent. Any material that aids in removing undesirable matter from the surface to be cleaned.

- **determinate growth**
a growth pattern in which first the vegetative phase of development is completed and then the plant enters the reproductive phase and all the seeds ripen approximately at the same time.

- **deuterium**
the isotope of the element hydrogen with one neutron and one proton in the nucleus, (atomic weight 2.0144). Designated D, d, or $H^z$.

- **deuteron**
a particle consisting of one proton and one neutron and having a positive charge of it is also the deuterium nucleus.

- **devaluation**
it means reducing the value of a nation's currency relative to gold or to the currency of other countries. It helps increase a country's physical exports and decreases its physical imports, provided other countries do not devalue their currencies.

- **deviation**
the spread or variability of the measures of a distribution with reference to any given point, usually, however, with reference to a measure of central tendency such as the mean or median.

- **deviation about the median (root mean square)**
the square root of the arithmetic mean of the squares of the deviations of values from their median.

- **dew**
water condensed onto plants and other objects near the ground usually over night.

- **dew point**
1. it is defined as the temperature at which the pressure exerted by the water vapour actually present in the air is the maximum for that temperature.
2. It may also be defined as the temperature at which the vapour actually present in the air is sufficient to saturate it.

- **dew point hygrometer**
a hygrometer for indicating the dew point from which the relative humidity can be calculated when the air temperature is known.

- **diagenesis**
transformations undergone by sediments after they have been deposited, under the effect of agents of external origin such as infiltrating water for e.g., consolidation of clay into schist, and of sand into sandstone.

- **diagnosis**
1. the identification of a pest, disease or deficiency of any nutrient.
2. the formal description of a plant, with special reference to the characters that distinguish it from related species.
3. identification of the nature and cause of a problem (soil or plant trouble).

- **diagnostic contacts to nonsoil material**
a contact is a root restricting layer, however, roots may penetrate the contact at infrequent intervals of greater than 10 cm.

- **diagnostic horizons**
combinations of specific soil characteristics that are indicative of certain classes of soils. Those which occur at the soil surface are called epipedon, those below the surface, diagnostic subsurface horizons.

- **dialysis**
the separation of colloids in solution from other dissolved substances by selective diffusion through a semi-permeable membrane. Such a membrane is slightly permeable to the molecules of the dissolved substances, but not to the larger molecules or groups of molecules in the colloidal state.

- **diameter law**
the rate of diffusion of a gas through a membrane is proportional to the mean diameter of the pores.

- **diamidine**
a molecule that has two amidine.

- **diamino**
a term used in chemical nomenclature to indicate the presence in a molecule of two amino groups.

- **diarylamine**
a molecule that contains an amine group and two aryl groups joined to the amino nitrogen.

**diatomaceous**
limnic materials that consist of diatomaceous earth.

**diatomaceous earth**
a geological deposit derived chiefly from the remains of diatoms.

**diatoms**
algae having a siliceous cell wall which persists as a skeleton after.

**dibasic**
compounds containing two hydrogens that may be replaced by a monovalent metal or radical.

**dibasic acid**
an acid having two hydrogen atoms capable of replacement by two basic atoms or radicals.

**dibble**
1. a pointed tool or sharp stick used to make holes in soil.
2. to sow or plant seed with a dibble.

**dibble to**
to sow seeds in shallow holes made with a suitable tool.

**dibbling**
a method of sowing seeds. It consists of sinking a shallow hole in the soil and dropping seed in it. The hole varies in depth depending upon the type of seed from 3 cm to 7 cm. Dibbling is done to maintain specific spacing and number of plants between the rows and within the rows.

**dicalcium phosphate**
is the salt of orthophosphoric acid. It is a by-product of the ammoniated superphosphate. It is formed by the contact of orthophosphoric acid and CaO in the rock phosphate
$H_3PO_4 + CaO \longrightarrow CaHPO_4 + H_2O$
Most of the $P_2O_5$ in dicalcium phosphate is citrate soluble but very little of it is water soluble. The importance of dicalcium phosphate lies in the fact that it could be a substitute for superphosphate when sulphur supplies are available to a limited extent.

**dicentric**
of a chromosome or chromatid having two centromeres.

**dichogamy**
the maturing of the anthers and ovules in the same flower at different times.

- **dicot weed**
weed producing two cotyledon seeds.

- **die off**
to die and fall to the ground, as a leaf.

- **dieback**
progressive death of shoots, branches, and roots generally starting at the tip.

- **dieldrin**
(HEOD : l,2,3,4,IO,10-hexachloro-6,7-epoxy-l,4,4a, 5,6,7,8,8a-octahydro-l,4,5,8-endo, exodimethanonaphthena) is a chlorinated hydrocarbon produced by the Diels-Alder reaction. It contains not less than 85% of the active ingredient HEOD.

- **differential host**
a particular species, or variety of plant whose reaction is used for determining physiological races of parasites.

- **differential thermal analysis (dta)**
the difference in temperature (AT) of the sample from that of an inert material is measured as a function of either time (t) or temperature (T°C), the sample being heated in an environment of uniformly increasing temperature. Different endothermic and exothermic changes are observed due to differences in structure or the ionic species with which the clay mineral are associated.

- **differential water rate**
it is a rate for the use of water taken on dry land from Government source of irrigation corresponding to the difference between the assessment of dry land and wet land.

- **differentiated products**
products which, in the eyes of consumers, have significant differences, because of either price, quality, advertising or other characteristics.

- **differentiating characteristic (soil taxonomy)**
a property chosen as the basis for grouping individuals (or classes of a lower level of generalisation) into groups. Individuals similar with respect to that characteristic are placed in the same group.

- **diffuse**
widely spread, not localised, not sharply defined at the margin.

- **diffuse layer**
a soil system which consists of a charged (negative) particle surface and an equal amount of counter ions (positive) ac-

cumulated in the liquid near the particle surface.

- **diffused runoff**
rainfall which runs off simultaneously over the whole area of the slope. Contrast to runoff concentrated to a single channel.

- **diffusion carrier**
a substance to which a dissolved particle becomes attached on one side of a membrane, moving with the dissolved particle through the membrane, and then depositing it on the other side.

- **diffusion coefficient (d)**
a parameter which can be measured and calculated, is an expression of the ion mobility. D has dimensions of length$^2$/time. The values of the diffusion coefficient for K vary from 2x $10^5$ cm$^2$/s in aqueous solution to about $10^6$ cm$^2$/s in a moist soil, $10^8$ cm$^2$/s in a dry soil, and to less than 1Q$^{-20}$ cm$^2$/s for interlayer potassium in unexpanded illitic clay. The average diffusion distance is $\sqrt{2Dt}$ in time t. Thus, potassium ions will move by diffusion 1 mm in about 4 h in a solution, but only 1 mm in about 16 years in illite.

- **diffusion constant**
the number of CC of $O_2$ or $CO_2$ gas which, having penetrated a porous body (for e.g., a soil), crosses 1 cm of thickness on 1 sq. cm. of surface in 1 second under a partial pressure of 1 mm mercury. It is proportional to the diffusion coefficient.

- **diffusion cutans**
are cutans having concentrations at a surface due to diffusion, they are usually associated with concentrations within the S-matrix of the soil material that reach a maximum at the cutanic surface.

- **diffusion nutrient**
a nutrient uptake mechanism. The movement of nutrient ions from the soil solution into root cells by passive or active absorption without mass flow of water. In passive absorption, each nutrient cation or anion moves independently from an area of high concentration of that nutrient to an area of low concentration of the same nutrient. In active absorption, ions move into cells against a concentration gradient using energy from respiration.

- **diffusion pressure**
  the pressure exerted by a diffusing substance, proportional to its concentration.
- **diffusion pressure deficit (DPD)**
  the amount by which the diffusion pressure of cell water or solution is lower than that of pure water at the same temperature and under atmospheric pressure. The DPD of a cell may be regarded as a measure of the pressure with which water would diffuse into the cell when immersed in pure water. The equation for DPD in a cell is:
  DPD=OP—TP
  Where OP is the osmotic pressure of the cell contents and TP is the turgor pressure within the cell. The DPD is a more sensitive indicator of the degree of turgidity.
- **diffusivity**
  the quantity of heat passing normally through a unit area per unit time divided by the product of specific heat, density, and temperature gradient.
- **digested sludge**
  sewage sludge which has undergone anaerobic sludge digestion or passed through an aerobic digester should have a fibrous structure and a peaty smell which is not unpleasant, unlike raw sludge, but it still may contain organisms that cause disease. Digested sludge is usually more difficult to dewater than raw sludge, although many texts state the opposite.
- **digester**
  in a waste water treatment plant, a closed tank that decreases the volume of solids and stabilises raw sludge by bacterial action.
- **digestible crude protein (DCP)**
  is the most common way of expressing the protein values and requirements for the ruminants in most of the countries. Digestible crude protein content of a feed stuff is obtained by multiplying the crude protein content by the digestibility coefficient of the protein in the fodder.
- **digestible energy**
  the apparent digestible energy of the food is the gross energy of the feed less the energy contained in faeces.
- **digestible nutrients**
  that portion of the nutrients consumed which are digested and taken, into the animal

body. The term is generally applied to the proteins, carbohydrates and fats.

- **digestion**
  the biochemical decomposition of organic matter. Digestion of sewage sludge takes place in tanks where the sludge decomposes, resulting in partial gasification, liquefaction,, and mineralisation of pollutants. Chemical breakdown of food into a form that can be assimilated by an organism.

- **diluents**
  1. materials used to dilute plant protection chemicals to secure coverage over wider area. Water is a common diluent. Finely ground inert materials like talc, clay, gypsum and lime are also used in diluting dusts.
  2. any liquid or solid material serving to dilute an active ingredient in the preparation of a formulation.

- **dilute**
  denoting a solution in which the amount of solute is low relative to that of the solvent.

- **diluvium**
  soil of slope bottoms, deep, fine-textured, rich in humus. It often covers buried natural soils.

- **dry land farming**
  1. the practice of crop production in low rainfall areas by conserving natural precipitation. Supplemental irrigation may be given from harvested run-off water.
  2. the practice of crop production without irrigation where rainfall is usually deficient.

- **dry lime sulphur**
  when lime sulphur concentrate is evaporated with sucrose as a stabiliser, probably the polysulphides are present as tetra and penta sulphides.

- **dry matter**
  the total amount of matter, as in a feed-stuff, less the moisture it contains.

- **dry matter per cent**
  the plant (or seed, fertiliser, soil, etc.) substance less water, found by oven-drying a weighed sample, weighing and determining per cent.

- **dry mix sulphur lime**
  in this form of lime sulphur, the temperature is allowed to rise due to slaking of calcium hydroxide. This is used on sulphur-sensitive plants.

- **dry nursery**
  a nursery that is maintained without any irrigation or other artificial watering.

- **dry processing of coffee**
sun drying of berries after removing the cover.

- **dry soil consistency**
the consistence of soil materials when dry is characterised by rigidity, brittleness, maximum resistance to pressure, more or less tendency to crush to a powder or to fragments with rather sharp edges, and inability of crushed material to cohere again when pressed together. To evaluate, select an air-dry mass and break in the hand, following classes are recognised:
a. *Loose:* noncoherent.
b. *Soft:* soil mass is very weakly coherent and fragile, breaks to powder or individual grains under very slight pressure.
c. *Slightly Hard:* weakly resistant to pressure, easily broken between thumb and forefinger.
d. *Hard:* moderately resistant to pressure, can be broken in the hands without difficulty but is barely breakable between thumb and forefinger.
e. *Very Hard:* very resistant to pressure, can be broken in the hands only with difficulty, not breakable between thumb and forefinger.
f. *Extremely Hard:* extremely resistant to pressure, cannot be broken in the hands.

- **dry spell**
1. a period, usually of not less than two weeks duration, during which no measurable rainfall occurs at a certain place or region. When extended into a month or more, a dry spell is considered a drought.
2. a period, of not less than four days, during which the daily maximum temperature remains at least 3 degrees to 4 degrees C above the normal maximum and the relative humidity continues below 50 per cent.

- **dry weather flow**
1. the flow of sewage in a sewer during dry weather. Such flow consists mainly of sewage and wastes, with no storm water included.
2. the flow of water in a stream during dry weather, usually contributed entirely by ground water.

- **dry weight**
weight of the product or materials after the removal of the moisture.

- **dry weight percentage**
the ratio of the weight of any constituent (of a soil) to the oven-dry weight of the soil.

- **dryer (drier)**
a unit which provides the conditions for removing moisture, generally by forced ventilation with or without the addition of heat, used in connection with conditioning of harvested produce.

- **drying**
the removal of moisture to a point which slows the rate of deterioration from chemical and biological activity, used in connection with conditioning of harvested produce.

- **drying front**
in a bin of drying grain, the advancing zone, which includes all of the grain that has just reached some arbitrarily, selected moisture content between the initial and final moisture content.

- **drying rate**
the rate at which the moisture is removed at any time from a product.

- **duck foot cultivation**
a field cultivator equipped with small sweep shovels.

- **duck root cultivator**
a field cultivator equipped with small sweep shovels.

- **duff**
acid humus, typically formed in free-draining acid soils of calcium concentration.

- **duma's method**
the nitrogen of an organic compound is estimated by heating organic compound with copper oxide. Then the combustion product is passed over a bright copper spiral, carbon, hydrogen and sulphur is oxidised to $CO_2$, water and $SO_2$ respectively. These can be absorbed in caustic potash solution. From the volume of nitrogen collected over caustic potash, percentage of nitrogen can be calculated.

- **dump raking**
pulling a rake behind a tractor to windrow or heap the broken vegetation for burning. Also can be used after cultural ploughing.

- **dumping**
means selling goods abroad at a lower price than is charged for them in the home market.

- **dune**
a mound or ridge of loose sand piled up by the wind, common where sand is abundant and wind usually strong along lake and sea shores and in some desert and semi-desert areas.

- **dune land**
consists of hills or ridges of

sand-sized particles drifted and piled up by the wind and either actively shifting or so recently fixed or stabilised that no soil horizons have developed. Active dune land is still drifting, and stabilised dune land is fixed by vegetation. Some areas mapped in the past as Stabilised dune land should be classed as a named soil. In places Blown-out land is associated with Dune land, although in mapping practice, transient areas of Blown-out land, developed in loose deep sand, are included in Sand-dune land, as are the areas of shifting sand that occur between dunes or on the slopes.

- **dung**
manure. The faeces or excrement of animals and birds.

- **duopoly**
it is a form of imperfect competition in which there are only two sellers of a commodity. Each seller under duopoly must consider the other firm's reaction to any changes that he may make in price or output. The interdependence of the two firms is great. There is a strong tendency for duopolist to reach a tacit agreement to fix a common price, limit the output, or divide the market on a territorial basis.

- **duopsony**
a market structure in which there are only two buyers for a commodity or a service.

- **duplicate gene**
genes which have an identical, but non-accumulative effect.

- **durable humus**
humus formed of neoformed colloidal complexes, much more resistant to microbial action compared to the humus of little decomposed organic matters. Intimately bound to the clay complexes.

- **duration curve**
a curve which expresses the relation of all the units of some item (for e.g., flow), arranged in order of magnitude as the ordinate, and time (frequently expressed in percentage) as the abscissa, A graphical representation of the number of times given quantities are equalled or exceeded during a certain period of record.

- **duricrust**
a hard surface layer in the soil that occurs chiefly in hot arid areas. It consists of various minerals, including iron and aluminium, usually strongly cemented by silica. It is belie-

ved that duricrusts are formed in areas of extensive flat terrain through capillarity, which draws strong solutions to the ground surface where they evaporate forming a hard deposit.

- **duripan**
an indurated horizon cemented with such materials as silica, calcium carbonate, or aluminium silicate. It is at least half cemented by $SiO_2$. Air dry peds do not slake in water or HC1 but will be destroyed by hot KOH after acid washing.

- **dust**
result of the crushing and sieving (0.2 to 0.5 mm) and artificial product comparable to sand fine-grained particulate matter that is capable of being suspended in air.

- **dust bowl**
a semiarid area in which the removal of vegetation for e.g., by drought, deforestation, or overgrazing) has resulted in extensive wind erosion of the top soil.

- **dust index**
the ratio, in per cent, of the quantity of matter passing through the US standard sieve number 200 (mesh of 0.074 mm) to that passing through the number 40 sieve (mesh of 0.42 mm).

- **dust mulch**
an induced loose, fine, granular or powdery condition in the surface soil, formerly believed to be effective in reducing evaporation losses of water from soil.

- **dust storm**
a storm in which dense masses of dust are carried by the wind. It is a common occurrence in desert or semi-desert areas where the surface deposits are loose and dry and the wind blows with sufficient force to raise the dust.

- **dusters**
plant protection equipments used to spread the chemicals available in the form of dust or powder on the plant foliage or on insect pests. The equipment throws out the powder in the form of a continuous cloud.

- **dusts**
these are solid form of pesticide containing a low concentration of the toxicant mixed with a carrier like talc.

- **dusty structure**
the structure of a soil in the state of small isolated juxtaposed grains.

- **dwarfing**
  the underdevelopment of a plant or plant organs. May be caused by any type of disease agent under certain conditions, or by faulty nutrition.

- **dynamic equilibrium**
  1. if two opposing processes are going on at the same rate in a system, thus keeping the system unchanged, the system is said to be in dynamic equilibrium, for e.g., a liquid in equilibrium with its saturated vapour, the rate of evaporation from the liquid surface is equal to the rate of condensation of the vapour.
  2. relates to a progressive economy. The incentive to change is to be found in all the organisms of the economy.

- **dynamics**
  is a method which takes into account all the changes, lags sequences, cumulative magnitudes and even expectations.

- **dynamometer**
  an instrument for measuring draft of tillage implements and for measuring resistance of soil to penetration by tillage implements.

- **dyne**
  the unit of force which, when acting upon a mass of 1 gram, will cause the mass to accelerate at the rate of 1 cm/sec$^2$.

- **dystrophic**
  said of a lake habitat in which the iron and humic acids in the water reduce the dissolved oxygen content.

- **EAC**
  suffix in the nomenclature of the taxonomic groups according to their tribes, as, for example, Streptococceae.

- **early maturity**
  the completion in a plant or animal of development of form and function sooner than others of the species.

- **early postemergence**
  after emergence of weeds but before crop emergence.

- **earth thermometers**
  are thermometers used to find the temperature at various depths in the soil. The earth thermometers are fixed on a frame which slides into an iron tube fixed in the ground.

- **ecad**
  a form modified by the conditions of its habitat (the modification not being heritable).

- **ecoclimate**
  the study of climate in relation to animal and plant life.

- **ecoclimatic adaptation**
  acclimatisation to the broad physical and climatic characters of any region.

- **ecocline**
  series of biotypes within a species that shows a genetic gradient correlated with a gradual environment gradient.

- **ecodesign**
  an expression of the cultural values of a given society which are attained through the application of technology to spatial forms that are adequate to the ecological characteristics of the environment.

- **ecodevelopment**
  a style of development which stresses specific solutions for the particular problems in each eco-region taking into account ecological and cultural contents as well as present and long term needs.

- **ecological niche**
  the status of an organism in the ecosystem, including its habitat and its effect on other organisms and on the environment.

- **ecological race**
  group of local populations within a species in which most individuals have similar environment and tolerances, wide ranging species may consist of many ecological races.

- **ecology**
  the science that deals with the study of the interrelationships of organisms in and to their environment.

- **ecomapping**
  the land use planning including inland waters, coastal areas and offshore.

- **econometrics**
  it is that branch of Economics which employs mathematical and statistical methods for the testing and verification of economic theories. The use of econometrics has been confined largely to problems which are included in partial equilibrium theory and in the theory of income and employment. Attempts at large-scale econometric models of an entire economy have not been very successful.

- **economic efficiency (irrigation)**
  the ratio of the actual income (net or gross income) attained with the operating irrigation system, compared with the income expected under ideal conditions. It is a measure of

overall efficiency, because it relates the final output to input.

- **economic friction**
  this takes place when non-economic factors hinder the smooth working of economic laws.
- **economic growth**
  means a sustained increase in per capita income accompanied by reduction in existing inequalities and economic betterment of the masses.
- **economic maturity**
  this term applies to a country which is economically fully developed.
- **economic model**
  it refers to mathematical statement of economic theory. It represents a method of analysis which present to us an over-simplified picture of the real economic world. As is well known, the real economic world is a very complex affair. Unless we eliminate the minor and unimportant factors, we cannot have a rigorous analysis of the real economic world. We can eliminate the minor and unimportant factors by making certain assumptions and, thus, set up an economic model on the basis of these assumption. Such a model can help us in analysing a specific problem in the light of these assumptions. An economic model is generally set out in an equation form with diagrams or with words.
- **economic nationalism**
  also known as autarchy. It refers to the objective of a nation to make itself economically independent of others.
- **economic paradoxes**
  the term refers to certain statements by economists which appear to be paradoxical. For example, an increase in saving on the part of some individuals during a period of depression may actually result in a reduction of the total saving of the community.
- **economic pest**
  a pest causing a crop loss of about 5-10%.
- **economic statistics**
  refers to that type of analysis where we establish the functional relation between two variables whose values relate to the same point of time or to the same period.
- **economic threshold**
  that point of pest infestation where application of a control

measure would return more money than the cost of the control procedure.

- **economic viability**
a scheme or project formulated should ultimately earn adequate return on the investment proposed, so as to enable the borrower to repay the loan along with the interest thereon leaving a reasonable surplus which can be used to maintain and/or increase the productivity of his farm or enterprise. The scheme or project can be considered as viable if the surplus available after meeting cultivation and normal family expenses will at least make it possible for the farmer to repay the loan with interest and maintain the productivity of the farm or enterprise.

- **economic yield**
part of the economic product from total dry matter.

- **economics of scale**
these refer to those economies which reduce the average production costs as the size of the plant increases. These economies can be discussed under two heads:
a) internal economies
b) external economies
The former results from the increased size of the individual firm while the latter results from the increased size of the industry as a whole. Internal economies (or, economies of mass production) result from the indivisibility of some factors and the advantage of specialisation. The advantage of internal economies are limited to a certain range of output only. It is the diseconomies of scale which will set in and offset the advantages accruing from internal economies. External economies of scale are those economies which no firm can gain through its own expansion but accrue to all the firms as the industry expands.

- **economy of higher wages**
it is believed by some people that higher wages will so increase the efficiency of the worker as to compensate the employer for the greater cost of labour.

- **ecophene**
the range of phenotypes produced by one genotype within a particular habitat. One genotype may be taken to mean a class of closely related genotypes.

- **ecospecies**
a category of variant individuals capable of genetic recom-

bination with other similar groups, without loss of fertility in the offsprings.

- **ecosystem**
  an ecological system, one in which there is a cycle of interchange of materials between living organisms and their environs. In a tropical rain forest, for example, the dead leaves and branches, built up by photosynthesis, drop to the ground, where they rot and provide the materials needed by the roots. The forest flourishes even though the soil is poor.

- **ecotone**
  when plant communities naturally merge into one another through a transition zone (do not change abruptly).

- **ecotope**
  The climatope and edaphotope of a place constitute the ecotope.

- **ecotype**
  1. an element of a Linnaean species, separable as being associated with particular habitat factors.
  2. a biotype resulting from selection with reference to a particular habitat.

- **ectodynamogenic soils**
  are soils which are dependent upon the climate.

- **ectodynamorphic soil**
  soil shaped by influences other than parent material.

- **ectophyte**
  a parasite growing on the surface of its host.

- **ectoplasm, ectoplast**
  the outer, clear cytoplasm laying against the cellwall.

- **ectotrophic mycorrhiza**
  a mycorrhizal association in which the fungal hyphae form a compact mantle on the surface of the roots. Mycelial strands extend inward between cortical cells and outward from the mantle to the surface soil.

- **$ed_{50}$ (effective dose 50)**
  dose of a pesticide that would prove lethal to 50% population of a test, pest species.

- **edaphic**
  1. pertaining to the soil.
  2. resulting from or influenced by factors inherent in the soil or other substrata, rather than by climatic factors.
  3. variations in soil texture, pH, presence of minerals and water which affect plant growth.

- **edaphic climax**
  a climax community which is determined by some property

of the soil.

- **edaphic factor**
any soil factor which affects the organisms-growing on or in it. Edaphic factors may be physical, chemical or biological and include pH, mineral and humus content etc.

- **edaphology**
soil science, particularly the influence of soil upon vegetation.

- **edaphon**
the organisms living within the soil, *i.e.* soil flora and fauna.

- **edible**
not toxic, nutritious, as a worm to a chicken, a chicken to a man, and a man to a worm.

- **edible oil**
any vegetable oil which is fit for humans to eat.

- **EDTA (ethylenediamine tetra acetic acid)**
a chelating agent which forms complex compounds with many cations. It is used to determine the concentration of calcium or magnesium ions, and thus the hardness of the water.

- **edulis**
a Latin word that means 'edible'.

- **effective demand**
demand for goods and services that can be made effective in the sense that the money is available to buy the goods and services in demand.

- **effective half life**
the half-life of a radioisotope in a biological system as a result of the combined effects of the biological half-life and the radiological half-life.

- **effective permeability**
the observed permeability of a porous medium to one fluid phase under conditions of physical interaction between this phase and other fluid phases present.

- **effective pore space**
the pore space through which the ions can migrate through easily. Larger the effective pore space, greater are chances of a nutrient ion becoming available to the plant. It is represented as a fraction of total pore space as shown below:
Pe= — x 100 P
Where P is equal to total pore space and p is the pore space through which ions can easily migrate, Pe the effective pore space.

- **effective porosity**
1. hydrology: often used in same sense as specific yield.

It is the ratio of the volume of water, oil, or other liquid, which after being saturated with that liquid, it will yield under any specified hydraulic conditions to its own volume.

2. the property of rock or soil containing intercommunicating interstices, expressed as a per cent of bulk volume occupied by such interstices.

- **effective rainfall**
precipitation falling during the growing period of the crop that is available to meet the evapotranspiration requirements of crops. It does not include precipitation lost through deep percolation below the root zone or through surface runoff.

- **effective rhizobium**
that strain of rhizobium which can induce nodulation and make nitrogen-fixation to occur on specific host plant.

- **effective root zone**
is the depth of soil from which the plants draw water and nutrients. It should not be confused with the true root zone.

- **effective size (soil mechanics).**
the maximum diameter of the smallest 10% of the particles of a sediment.

- **effective size of grain (hydrology)**
the diameter of the grains in an assumed rock or soil that would transmit water at the same rate as the rock or soil under consideration but that is composed of spherical grains of equal size and arranged in a specified manner.

- **effervescence**
the bubbling of solution of an element or chemical compound as the result of the emission of gas without the application of heat, for example, the escape of carbon dioxide from carbonated water.

- **efficiency index**
the rate at which dry matter accumulates in plants.

- **efficiency of a machine**
the ratio of the output energy to the input energy. The efficiency of a machine can never be greater than unity. Often expressed as a percentage.

- **efficiency of irrigation**
the fraction of the water diverted from a river or other source that is consumed by the crop. Often applied to whole irrigation systems and takes account of conveyance losses.

- **efficiency of water application**

  the fraction of the water delivered to the farm that is stored in the root zone for use by the crop.

- **efflorescence**

  the phenomenon whereby crystals lose their water of crystallisation when exposed to air at ordinary temperatures is known as efflorescence and the crystals are said to be efflorescent.

- **effluent**

  a discharge of pollutants into the environment, partially or completely treated or in its natural state. Generally used in regard to discharges into waters:

  1. flowing forth out of. Geology: Flowing out, as lava through fissures in the side of a volcano, as a river from a lake.

  2. anything that flows forth. Geography: a stream flowing out of another or forming the outlet of a lake.

  3. a discharge of pollutants into the environment, partially or completely treated in its natural state. Generally used in regard to discharges into waters.

  4. partially or completely treated sewage flowing out of any sewage treatment device.

- **effusion (chemistry of gases)**

  the passage of gases through small apertures under pressure. The relative rates of effusion of different gases under the same conditions are inversely proportional to the square roots of their densities.

- **eichler's rule**

  the groups of hosts with more variation are parasitised by more species than taxonomically uniform groups.

- **eisodal aperture**

  the enlargement of the stomatal pore nearest to the surface of the leaf.

- **ektodynamorphous soil**

  a soil whose characteristics are primarily affected by external factors.

- **elaioplast**

  a plastid which contains stored oil.

- **elastic collision**

  a collision between particles in which there is no change in total kinetic energy of the particles and/or quanta.

- **elasticity of demand**
the degree of responsiveness of demand to changes in price.

- **elasticity of substitution**
the extent to which one commodity can be substituted for the other. Perfect elasticity of substitution would occur if the two commodities were perfect substitutes of each other.

- **elasticity of supply**
the degree of responsiveness of supply to changes in price.

- **elcctrovalence**
the valence of an atom that has formed an ionic bond.

- **electric double layer**
a phenomenon found at a solid-liquid interface, it is made up of ions of one charge type which are fixed to the surface of the solid and an equal number of mobile ions of the opposite charge which are distributed through the neighbouring region of the liquid, in such a system the movement of liquid causes a displacement of the mobile counterions with respect to the fixed charges on the solid surface.

- **electric furnace slag**
is a by-product resulting from the electric furnace reduction of rock phosphate in the preparation of elemental P. It is produced due to fusion of silica and calcium oxide which is then drawn off and quenched with water.
It is largely calcium silicate having a $CaCO_3$ equivalence of 60—80%. It also contains 0.9 to 2.3% PaOs.

- **electric logging**
it consists of a record of the apparent resistivities of the subsurface formations and the spontaneous potentials generated in the bore-hole, both plotted in terms of depth below the ground surface. These .two properties are related indirectly to the character of the subsurface formations and to the quality of water contained in them.

- **electric potential (v/e)**
it results from the electrical field to which water may be subjected during laboratory and field investigations. It forms a part of total soil water potential.

- **electric resistance blocks**
the dependence of electrical conductivity of a porous solid on the amount of water it contains, forms the basis of this method of determining

water content of soils. The resistance blocks are made of various materials like fibreglass, gypsum, nylon etc. When embedded in soil the moisture content of a porous block containing electrodes comes in equilibrium with the moisture of surrounding soil. The electrical resistance between the two electrodes varies with moisture of porous block which is in equilibrium with soil moisture. A resistance meter measures the resistance. The blocks are calibrated against a range of moisture and are installed in the field at required depths. The electrode lead wires are taken out of the soil surface and are well secured and protected against rain. When required the resistance meter is carried to the field and through the electrode leads the resistance is measured to know the moisture content from the calibration curve. These blocks are not useful for saline soils, for the resistance reading is affected by salt concentration.

- **electroculture**
the stimulation of growth of plants by electrical means.

- **electrodialysis**
dialysis that is conducted with the aid of an electromotive force applied to electrodes adjacent to both sides of the membrane.

- **electrokinetic phenomena**
the phenomena associated with movement of charged particles through a continuous medium or with the movement of a continuous medium over a charged surface.

- **electrokinetic potential**
1. in a colloidal system, the difference in potential between the immovable layer attached to the surface of the dispersed phase and the dispersion medium
2. The work done in bringing a unit charge from infinity to a reference point in the liquid layer attached to the solid phase in a colloidal system.

- **elemental sulphur**
is a yellow powder ranging in purity from 50 per cent to more than 99 per cent. It is not soluble in water and does not supply calcium or $SO_4$-S directly. When applied to soils, sulphur is slowly oxidised by micro-organisms in the presence of air and water to sulphuric acid. If lime is present in the soil, it reacts with the sulphuric acid to

form gypsum. Finely ground sulphur will react more quickly in the soil than will coarsely ground material, but the latter is easier to apply and is less irritating to the skin and eyes.

- **elite variety**
an improved variety developed by plant breeders and released to farmers because of its superiority in at least one respect,

- **elongation stage**
the period of growth when an organ increases in length, but before secondary thickening begins.

- **elute**
to wash off, or remove from, specifically, to remove by extraction a chemical compound from a substance on which it has been adsorbed.

- **elution**
purification by washing and pouring off the lighter matter suspended in water, leaving the heavier portions behind.

- **elutriation**
the washing, separation, or sizing of fine particles of different weight by suspending them in a current of air or water.

- **elutriation method**
a type of mechanical analysis of soil where the separation of different soil fraction is effected by washing out particles in a rising current of water (particles whose settling velocity is equal to or less than the velocity of the rising water will be taken away by the current).

- **eluvial horizon**
a soil horizon from which fine material has been removed either in solution or in water suspension.

- **eluviation**
the removal of material in solution or of finer fractions in suspension from surface to the lower layers of soil profile with percolating water.

- **emasculation**
removal of stamens from hermaphrodite flowers before they have liberated pollen, as a preliminary to artificial hybridisation.

- **embedded grain cutans**
are those on coarse grains that, along with the cutans, are embedded in the plasmic matrix.

- **emergence**
the action of a young plant breaking through the surface

of the soil, or of an insect coming out of an egg or pupa.

- **emerson enhancement effect**
the sum of photosynthetic activity caused by short wavelength light and long wavelength light, is less than the activity produced by simultaneous radiation by both types of light.

- **eminent domain**
it refers to the right of a government to take private property for public use even against the owner's will, provided that just compensation is paid.

- **emissive power of the soil**
the power of the soil to emit heat.

- **employment function**
this term was used by Lord Keynes. It is the mathematical relationship between effective demand, as measured in terms of the wage-unit, and the amount of employment.

- **employment, secondary**
in addition to the primary employment, a given investment will also create secondary employment consequent upon the increased spending of the workers who are drawn into primary employment.

- **emulsify**
to make into an emulsion, when small drops of one liquid are finely dispersed (distributed) in another liquid, an emulsion is formed. The drops are held in suspension by an emulsifying agent, which surrounds each drop and makes a coating around it.

- **emulsifying agent**
a material which facilitates the suspending of one liquid in another, for example, oil dispersed in water.

- **en masse erosion**
mechanical erosion which removes materials of all sizes.

- **encapsulation**
a method of formulating pesticides, in which the active ingredient is encased in a material (often poly vinyl) resulting in sustained pesticidal release and decreased hazard. Also, a method of disposal of pesticides and pesticide containers by sealing them in sturdy, waterproof, chemical-proof containers which are then sealed in thick plastic, steel or concrete to resist damage of breakage so the contents cannot get out.

- **encrustation**
the act of encrusting, of form-

ing an adherent crust on the surface of an object.

- **encrusting water**
  Water which deposits calcium carbonate.

- **end contraction**
  the horizontal distance from the ends of the weir crest to the sides of the weir pond.

- **end point**
  a stage in the titration at which an effect, such as a colour change occurs, indicating that a desired point in the titration has been reached.

- **endodynamorphic soil**
  soil influenced mainly by parent material.

- **endodynamorphic soil**
  a soil whose characteristics are mainly determined by a parent material.

- **endoenzyme**
  an enzyme formed within the cell and not excreted into the medium. It is also known intracellular enzyme.

- **endoergic reaction**
  a nuclear reaction which absorbs energy.

- **endogenous**
  growing or originating from within.

- **endogenous business cycle theory**
  this theory brings out the self-generating nature of the business cycle. The psychological, monetary and credit monetary over investment and under-consumption theories are the important endogenous theories.

- **endomitosis**
  the doubling of the chromosomes without the division of the nucleus, resulting in polyploidy.

- **endophyte**
  a plant living inside another plant, but not a parasite.

- **endos**
  a set of two bands of earth turned over by the plough and inclined to one another.

- **endossement**
  a method of tillage in which the furrows form a field on which half the furrows are inclined in one direction and the other half in the other.

- **endosulphan**
  (thiodan: 6, 7, 8, 9, 10, 10-hexachloro-l, 5, 5a, 6, 9, 9a-hexahydro-6, 9-methano-2, 4, 3-benzpdioxathin-3-oxide) is a chlorinated hydro-carbon containing 4 parts of X-cia isomer and 1 part of B-trans isomer.

- **endotoxins**
toxin that remain within the cells that produce them and do not actively stimulate the production of corresponding antitoxins.

- **endotrophic**
nourished or receiving nourishment from within, as fungi or their hyphae receiving nourishment from plant roots in a mycorrhizal association.

- **endotrophic mycorrhiza**
a mycorrhizal association in which the fungal hyphae are present on the root surface only as individual threads that may penetrate directly into root hairs, other epidermal cells, and occasion into cortical cells. Individual threads extend from the root surface outward into the surrounding soil.

- **energid**
a volume of protoplasm which is controlled by the enclosed nucleus.

- **energy budget**
the balance between energy inputs and energy demands in an ecosystem.

- **energy chain**
a sequence of energy-transforming systems each of which receives energy from the preceding system and supplies it to the next.

- **energy conservation law**
energy flowing into a system is equal to energy stored in the system plus energy flowing out of it (all measured in heat equivalents).

- **energy degradation law**
whether stored or being used, concentrations of energy spontaneously disperse, losing their potential for doing work.

- **energy flow**
trapping of solar energy by photosynthetic organisms and its transformation into chemical energy and passage through various trophic levels in an ecosystem.

- **energy source**
the source from which an organism derives the energy for metabolic activities, *e.g.* sunlight, sulphur, cellulose, hydrogen, etc.

- **energy value**
the amount of heat each gram gives out on complete combustion.

- **engel's law**
according to this law, the smaller a person's income, the greater the proportion of it

that he will spend on food.

- **enlargement**
primary growth in thickness, before secondary thickening.

- **enrichment**
the process of increasing the proportion of one isotope in a given mixture.

- **enrichment (soil)**
it may be with respect to horizons but is usually taken in the sense of the entire soil receiving material from surrounding pedons as in depressional parts of the landscape or by air from remote areas.

- **enrichment culture**
a technique in which environmental (including nutritional) conditions are controlled to favour the development of a specific organism or group of organisms.

- **environment**
the complete range of external conditions under which an organism lives, including physical, chemical and biological factors, such as temperature, light, and the availability of food and water.

- **enzyme**
a protein that accelerates (catalyses) specific transformations of material as in the digestion of foods.

- **enzyme gene theory**
the theory that a gene acts by controlling the synthesis of one enzyme. A single gene mutation will result in the blocking of a single step in a metabolic pathway and consequently will inhibit the completion of a chain of reactions.

- **enzymic**
pertaining to the action of an enzyme.

- **ephemer**
an introduced plant which is unable to survive and soon disappears.

- **ephemeral**
a plant which completes its life cycle in a very short time.

- **ephiphytotic**
an epidemic among plants.

- **epicotyl**
that portion of seedling-stem which is above cotyledons.

- **epictesis**
the ability of living cells to accumulate salts in a higher concentration than that in which they occur in the surrounding solution.

- **epidemic**
the unusual prevalence or sudden appearance of a given infectious disease for a limited period of time affecting a high

proportion of the human population. The analogous situation among plants and animals is referred to, as Epiphytotic and epizootic respectively.

- **epidemiology**
the science of tracing the sources from which diseases spread.

- **epimers**
the sugars which differ from one another only in the configuration around a single carbon atom are known as epimers. *e.g.*, galactose and glucose differ with respect to carbon 4 while glucose and mannose differ with respect to carbon 2 only.

- **epinasty**
the twisting or curling of leaves and stems caused by uneven growth of cells, especially in leaves, in which the upper surface grows faster than the lower surface and thus causes the leaf edges to bend down.

- **epipedons**
are simply the uppermost soil horizons. Epipedon is not synonymous with a horizon and may be thinner than the A horizon or include some of the B horizon.

- **epiphyte**
any plant growing upon or attached to another plant or object merely for physical support. They are often known as air plants because they are not attached to the ground, but obtain water and minerals from the rain and from debris that fall on the support, *e.g.* ferns, orchids, mosses, lichens and algae.

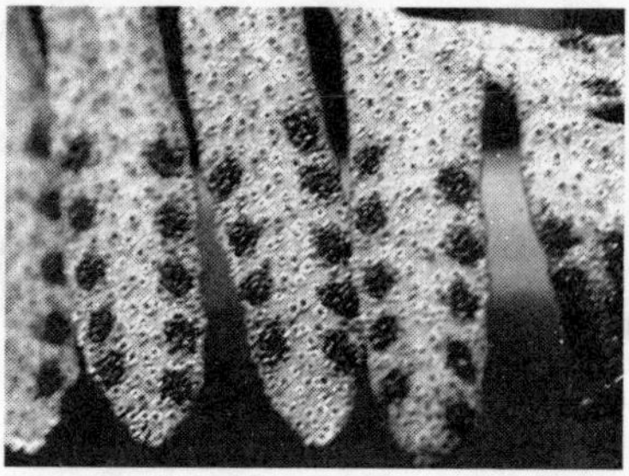

- **epiphytotic**
a widespread occurrence of a plant disease.

- **epitrophic**
1. having buds on the upper side.
2. growing more on the upper than lower side.

- **EPM (equivalent per million)**
a unit chemical equivalent weight of salt per million unit weight of soil or solution.

- **equal product curve**
a curve showing the same amount of product obtained by the employment of differ-

ent quantities of two factors, say X and F.

- **equality of saving and investment**
the concept was put forward by Lord Keynes in his magnum opus 'The General Theory' in 1936. Before that Lord Keynes did not believe in the equality of saving and investment. According to him, it was inequality between saving and investment which led to the operation of the business cycle. When saving exceeded investment, it was believed that a depression would occur, when investment exceeded saving it was thought that inflationary boom would set in. However, in his General Theory, Keynes discarded this belief and held that saving and investment were always equal to each other.

- **equilibrium**
it denotes a situation in which the various factors involved are in such a degree of balance that they tend to result in reasonable stability under conditions existing at a particular time.

- **equilibrium moisture content**
the moisture content of the product when it is in equilibrium with the moisture content of the surroundings or hygroscopic equilibrium.

- **equilibrium phosphate potential (EPP)**
is determined in a phosphate solution of such a concentration that there was no loss or gain of P when it was shaken with the soil. The EPP was found to be a better index of the phosphorus available to plants compared to phosphate potential.

- **equitorial plate**
the area across the centre of the spindle where the chromosomes become attached at metaphase.

- **equivalent acidity**
a term used to express the calcium carbonate equivalent of the acidic residue left by a fertiliser material.

- **equivalent basicity**
the basic residue to any fertiliser measured in terms of calcium carbonate required to neutralise it. Term for expressing the basicity of a fertiliser.

- **equivalent diameter**
the diameter of a hypothetical sphere composed of material having the same specific

gravity as that of the actual soil particle and of such size that it will settle in a given liquid at the same terminal velocity as the actual soil particle.

- **equivalent diffusion distance**

it is the length of a tube of unit cross-section area through which free diffusion will supply oxygen to a given point at the same rate as it is supplied by the soil. Equivalent diffusion distance is taken equal to linear distance which is not so because of tortuosity in soil pores.

- **equivalent field moisture**

the minimum water content, expressed as a percentage of the weight of the oven dried soil at which a drop of water placed on a smoothed surface of the soil will not immediately be absorbed by the soil but will spread out over the surface and give it a shiny appearance.

- **equivalent fluid**

a hypothetical fluid having a unit weight such that it will produce a pressure against a lateral support presumed to be equivalent to that produced by the actual solid.

- **equivalent per million (EPM)**

the concentration of a substance in mg/litre, divided by its equivalent weight. This is the same as its miliiequivalent per litre if the solution has a specific gravity of 1.

- **equivalent radius**

a measure of sedimentary particle size equal to the radius of a spherical particle of the same density that would have the same settling rate as the sedimentary particles.

- **equivalent spherical diameter**

1. an experimentally accessible characteristic of small particles of irregular shape, the radius of a spherical particle that would sediment with the same velocity as the non-spherical particle actually observed.
2. Used to define particle size fractions in soil suspensions.

- **equivalent weight**

the number of parts by weight of an element or compound which will combine with or replace, directly or indirectly, 1.008 parts by weight of hydrogen, 8.00 parts of oxygen, or the equivalent weight of any other element or compound.

- **eradicant**
  1. a chemical used to eliminate a pest or pathogen from a plant or a place in the environment.
  2. any chemical or compound used for destroying plants.

- **eradication**
  1. a method of disease control in which the pathogen is eliminated after it has established.
  2. absolute destruction, by burning or deep ploughing etc.

- **erg**
  the unit of work done by a force of 1 dyne acting through a distance of 1 cm.

- **ergodic hypothesis**
  the succession of vegetation in different places at one time will show the same sequence that one place will show in successive times.

- **ergosterol**
  a sterol which upon exposure to ultra-violet light, acquires the properties of vitamin D.

- **ergot**
  a fungal disease of cereals, particularly rye, and grasses caused by the parasite *Claviceps purpurea* of the *Ascomycetes*. Dark-purple ergot 'grains' replace the cereal grains of the head. The ergot 'grains' are dense masses of cells of the fungus.

- **erio**
  a Greek prefix meaning 'woolly'.

- **erode to**
  to wear away or remove the land surface by wind, water or other agents.

- **erodibility**
  the relative ease with which one soil erodes under specified conditions of slope as compared with other soils under the same conditions, this applies to both sheet and gully erosion.

- **erodible**
  susceptible to erosion, *e.g.* a soil that is quite susceptible to erosion is referred to as erodible, while one that is resistant to erosion is said to be relatively non-erodible. Soils may generally be referred to as highly erodible, moderately erodible, slightly erodible, etc.

- **erosion**
  1. the wearing away of the land surface by running water, wind, or other geological

agents, including such processes as gravitational creep.
2. all processes by which earthly materials or rocks are loosened and moved from place to place.
3. detachment and movement of soil or rock material by water, wind, ice or gravity.

- **erosion accelerated**
erosion much more rapid than normal or natural geological erosion, primarily as a result of the influence of the activities of man or, in some cases, of other animals or natural catastrophies that expose base surfaces, for example, fires.

- **erosion control**
the application of necessary measures to control accelerated erosion of land surfaces by vegetative or by artificial structures such as terraces, dams, retards, baffles, etc., used to control erosion, in contrast to vegetative control.

- **erosion gully**
the erosion process whereby water accumulates in narrow channels and, over short periods, removes the sail from this narrow area to considerable depths, ranging from less than a meter to 30 meters or more.

- **erosion normal**
the removal of soil and rock material by water, wind and gravity, in a manner and at a rate characteristic of the land surface in its natural environment, sometimes referred to as geological erosion.

- **erosion pavement**
a layer of coarse fragments, such as sand or gravel, remaining on the surface of the ground after the removal of fine particles by erosion.

- **erosion resistant crop**
a crop, because of its dense foliage, root system etc. providing effective protection against soil loss by erosion.

- **erosion rill**
a form of erosion caused by water, in which numerous very small and more or less

straight channels are produced, the channels get obliterated by ordinary use of the land.

- **erosion sheet**
the erosion of a fairly uniform layer of material from the land surface, often scarcely perceptible over short periods, especially when caused by wind.

- **erosion splash**
the shattering of small soil particles caused *by* the impact of raindrops on wet soils. The loosened and shattered particles may or may not be subsequently removed by surface run-off.

- **erosive**
tending to cause erosion, the term applies to the eroding agent, such as wind or water. The terms has also been used interchangeably with 'credible', but the word 'erosive' is preferred when referring to the agent causing erosion, and 'credible' when referring to the material eroded.

- **error**
the observed or obtained value of a measurement of a quantity minus the true value or mean value.

- **error experimental**
the variation between plots or other experimental units, due to causes other than the treatments that have been applied. May refer to a variance or a standard error.

- **error of measurement**
the amount by which a given measurement or observation deviated from the theoretical true value, or from the most probable value, or from an estimate of the most probable value, of the quantity measured. Also called error of observation.

- **error of prediction**
the amount by which the actual value of an observation differs from an estimated or computed value, or the deviation of a value of a dependent variable from the regression line.

- **erythrophyll**
a red anthocyanin as found in some leaves and red algae.

- **escape**
an accidental plant introduction, which has escaped from cultivation.

- **essence**
a substance that contains a high degree of essential oils or other stored products of a plant.

**essential element**
a chemical element, without which a plant cannot develop, and will ultimately die.

**essential fatty acids**
three unsaturated fatty acids, lino-lenic acid ($C_{17}H_{29}$ COOH), linolenic acid ($C_{17}H_3$iCOOH) and arachidonic acid ($Ci_9H_3$iCOOH), that are essential for the growth of mammals. As constituents of phospholipids and glycerides, they are vital for membrane production and fat metabolism. Only linoleic acid is required in the diet as the other two fatty acids can be synthesised from it.

**establishment**
development of a new crop, naturally or assisted, to a stage where the young regeneration, natural or artificial, is considered safe from normal adverse influences such as frost, drought, or weeds, and no longer needs special protection or tending operations other than cleaning, thinning and pruning.

**estcrification**
the fatty acids react with alcohols in the presence of dehydrating agent like anhydrous zinc chloride or concentrated sulphuric acid to form esters.

**ester sulphate**
also called hi-reducible sulphur. It is more precisely defined as that fraction of organic sulphur that is reduced to $H_2S$ by a mixture of hydriodic, formic and hypophosphorus acids. This mixture produces $H_2S$ only from compounds containing the C-O—S linkage (ester sulphate), C—N—S linkage (sulphamate), and from some organic sulphites such as dimethyl sulphite. The C-S bond is not ruptured under these conditions.

**estimate**
to use any statistic derived from a sample to infer the value of a parameter in the universe from which the sample was taken or the value of a statistic derived from another sample.

**estimate of error**
an objective estimate of the magnitude of the experimental error, commonly consists of the standard error or error variance.

**estivation**
reduction of biologic activity by an organism during the

summer, more generally, during periods that are hot, or dry, or both.

- **ethylene**
  a gas $C_2H_4$ which is produced by plants in minute amounts and has various developmental effects as a hormone, including regulation of fruit ripening.

- **etiolation**
  the type of growth exhibited by plants grown in darkness, usually from seed. They lack chlorophyll and therefore appear white or yellow. They show less differentiation and contain reduced amounts of supporting material, such as lignin. They are therefore more fragile. Light restores normal growth.

- **eu**
  a prefix from the Greek meaning well, good, etc., applied in microbiology to mean true as in eubacteria.

- **eubacteria**
  a large and diverse group of bacteria, principally distinguishable by their rigid cell walls that give the cells a fixed form. Most are unicells that divide by binary fission. The cells can be spherical, rod-shaped, or helical. Most are immo-tile, but some possess flagella. They are a ubiquitous group, some being found in extreme conditions.

- **eucarpic**
  forming reproductive structures on certain portions of the thallus, the thallus itself continuing to perform its somatic functions.

- **euler's theorem**
  according to this theorem, if a production function involves constant returns to scales, the sum of the marginal products will be equal to the total product. This theorem has played significant role in the development of the marginal productivity theory of distribution.

- **euphotic zone**
  that part of the water which receives sufficient sunlight for plants to be able to photosynthesise. The zone in water where there is sufficient light for algal productivity. Surface layer of water to the depth of light penetration at which photosynthesis of organisms just balances their respiration. The layer of water, which receives ample sunlight for photosynthesis. It varies with the light extinction coefficient. Usually no deeper than about 60 m.

■ **euphotometric**
said of a leaf which is permanently placed in a plane perpendicular to the direction of the strongest diffuse light reaching it.

■ **euryhaline**
tolerant of considerable difference in salinity, generally refers to marine organisms.

■ **eurythermous**
tolerant of a wide range of depths temperatures.

■ **eutrophic**
having concentration of nutrients optimal or nearly optimal for plant or animal growth.

■ **eutrophic lakes**
shallow lakes, weed chocked at the edges and very rich in nutrients. The water is characterised by large amounts of algae, low water transparency, low dissolved oxygen, and high bod. A shallow lake that has good primary productivity, usually abundant littoral vegetation, and dense plantcton population. Blooms occur when nutrient and physical conditions are right. During later stages of entrophication the bottom water tends to become depleted of oxygen during the summer.

■ **eutrophic waters**
with a good supply of nutrients, they may support rich organic production, such as algal blooms.

■ **eutrophication**
the normally slow aging process by which a lake evolves into a bog or marsh and ultimately assumes a completely terrestrial state and disappears. During eutrophication the lake becomes so rich in nutritive compounds, especially nitrogen and phosphorus, that algae and other microscopic plant life becomes superabundant, thereby 'choking' the lake and causing it eventually to dry up. Eutrophication may be .accelerated by many human activities. The enrichment of a body of water with nutrients, the intentional or unintentional enrichment of water: Enrichment of bodies of water, primarily caused by sewage and runoff from fertilised agricultural land.

■ **evaluation**
a process of judging the performance of the system under examination. In the context of project evaluation it is a process of which the project

activities and results are studied, analysed and judged against certain explicitly established criteria. These criteria are usually related to the project objective.

- **evaporation (e)**
the process during which a liquid changes into a gas. The process of evaporation of water in nature is one of the basic components of the hydrological cycle by which water changes to vapour through the absorption of heat energy. The essential requirements for evaporation process are the source of heat to vaporise the liquid water, and the presence of a concentration gradient of water vapour between the evaporating surface and the surrounding air.

- **evaporation ponds**
shallow, artificial ponds where sewage sludge is pumped, permitted to dry, and either removed or buried by more sludge.

- **evaporation potential**
the amount of water evaporated per unit of time from a free water surface. A theoretical value greater than actual evaporation.

- **evaporation power**
an index of the degree to which a region is favourable or unfavourable to evaporation. Refers to a unit area exposed parallel to the wind and expressed in the same terms, it is equivalent to evaporative capacity or evaporability.

- **evaporation rate**
the quantity of water, expressed in terms of depth of liquid water, which is evaporated from a given water surface per unit of time. It is usually expressed in inches depth per day, month, or year.

- **evaporation station**
a station set up on the surface of body of water or on land for making evaporation measurements.

- **evaporativity**
the potential rate of evaporation as distinguished from the actual rate. The rate of evaporation under existing atmospheric conditions from the surface of a body of water that is chemically pure and that has the temperature of the atmosphere. It is generally expressed in terms of the depth of water removed from such surface in a given unit of time. A more accurate method of expressing it, used by plant

physiologists is in terms of weight units of chemically pure water lost per unit of time from some specified standard evaporating surface.

- **evaporimeter**
  is an instrument for measuring the amount and rate of evaporation.

- **evapotranspiration (ET)**
  the combined processes of evaporation of water from a soil plus the transpiration through plants growing on it.

- **evergreen**
  1. vascular plants which do not shed all their leaves at the same time and so appear green all the year round.
  2. never entirely without green foliage, leaves persisting until a new set has appeared.

- **evergreen forest**
  forest composed chiefly of evergreen dominants.

- **ex ante evaluation**
  a pre-project evaluation to determine the project's feasibility and desirability.

- **ex post evaluation**
  an evaluation made after a project has been completed, to find out whether the project contains any deficiency.

- **exacting**
  making high demands as regards soil and other factors.

- **excess water**
  water, particularly rain water, over and above what is needed for plant growth.

- **excessively drained soils**
  soils that are very porous and freely permeable to great depths. No mottling occurs in any horizon, indicating that no artificial drainage is needed.

- **exchange acidity**
  the titrable hydrogen and aluminium that can be replaced from the adsorption complex by a neutral salt solution. Usually expressed as milliequivalents per 100 grams of soil.

- **exchange reaction**
  the transfer of atoms, radioactive or stable, of the same element, between molecules, or from one side to another on the same molecule.

- **exchangeable ions**
  those ions held on the soil complex that may be replaced by other ions. There are ions held so tightly that they cannot be exchanged.

- **exchangeable phosphate**
  the phosphate anion reversibly attached to the surface of the

solid phase of the soil in such form that it may go into solution by anionic equilibrium reaction with isotopes of phosphorus or with other anions of the liquid phase without solution of the colloid phase to which it was attached.

- **exchangeable potassium**
the potassium which is held mainly by the colloidal portion of the soil and which is easily exchanged with the cations of neutral salt solutions. It is readily available to growing plants.

- **exchangeable sodium**
is that fraction of the soil sodium that is adsorbed by the soil and may be replaced by, or exchanged with, other salt constituents, such as calcium or magnesium.

- **exchangeable sodium percentage (ESP)**
the degree of saturation of the soil exchange complex with sodium. It may be calculated by the formula: ESP=exchangeable sodium (me/ 100 gm soil) divided by cation-exchange capacity (me/100 gm soil) times 100.

- **excited state**
a state of an atom, molecule or other system with an energy greater than that of the ground state.

- **excreta**
waste material eliminated from body or any tissue thereof, deleterious substances formed within a plant.

- **excretions**
substances that have become so changed in composition through metabolism that they are no longer useful to the organism that produced them and are cast off, e.g. carbon dioxide.

- **exine**
the outer layer of a spore wall.

- **exodermis**
the outer layer of subrized cells on a root. It replaces the piliferous layer in the older parts of the root.

- **exoergic reaction**
a nuclear reaction which liberates energy.

- **exogamy**
the fusion of gametes produced by organisms that are not closely related.

- **exotic**
not native to a given area.

- **exotic plant**
an introduced plant not fully naturalised or acclimated.

■ **exotoxins**
toxins that diffuse from the cells that produce them into the surrounding medium. They are antigenic and stimulate the formation of antitoxins.

■ **expansion path**
the locus of least cost combinations of $X^*$ and $X_y$ for producing varying levels of output. It is essentially a long run concept.

■ **expectational equilibrium**
a condition in which the expectations of the various organisms of society are mutually compatible and are fulfilled.

■ **f**
1. the suffix f is used for soil that is thought to be permanently frozen.
2. a symbol for the plot which has been fertilised, employed in systematic fertiliser trials.

■ **facilitated diffusion**
transport of molecules or ions along a concentration gradient by a carrier system, but not requiring expenditure of energy.

■ **factor**
a circumstance that affects the result of an observation or experiment. Used especially of a set of related treatments in an experiment, each treatment of the set is a different state or level of the factor, *e.g.* different varieties of the same plant species, different levels of application of a fertiliser, or different methods of cultivation.

■ **factor cost**
this refers to the price of a commodity paid by the consumer minus any tax or duty included in its price. Actually, this amount accrues to the factors of production which take part in the production of a commodity. The national income of a country is calculated at factor cost, otherwise the national income of one year could not be compared with that of the other, as an increase in indirect taxation would falsely raise the amount of national income.

■ **factor of the habitat**
anything in the environment which affects, directly or indirectly the life of a plant.

■ **factorial design**
the plan of the experiment that permits the simultaneous

evaluation of more than one experimental factor, as contrasted with the single variable design.

- **factorial experiment**

an experiment in which all the treatments or agents under investigation are varied simultaneously, and combined in such a way that any derived effect of one or group of them may be isolated, and evaluated separately.

- **factors of soil formation**

a factor of soil formation is an agent, force, condition, or relationship, or combination of these, which influences, has influenced, or may influence a parent material of soil, with the potential of changing it. Dokuchaev gave the following equation of the factors:

S=f (Cl, O, P)t°

where S represents soil, Cl climate of a region, O the organisms (both plants and animals), P the geological substratum, and t° the relative age of soil (young, mature, or senile).

- **facultative**

1. voluntary, optional, having a power to do or not to do.
2. having the power of living under different conditions.

- **facultative anaerobe**

an organism that can respire under aerobic and anaerobic conditions.

- **facultative parasite**

1. an organism capable of infecting another living organism or of growing on dead organic matter, according to circumstances.
2. organisms that can live either as parasites or as saprophytes.

- **facultative saprobe**

an organism capable of growing on dead organic matter, or of infecting another living organism, according to circumstances.

- **facultatively anaerobic**

organisms that can grow in either the presence or absence of air.

- **fahrenheit scale**

a scale in which the normal freezing point of water is 32° and the normal boiling point of water is 212°. Since there are 100° Celsius degree and 180 Fahrenheit degree between the freezing and boiling point of water, 5 Celsius degrees equal 9 Fahrenheit degrees.

- **fairy ring**
  a ring of mushrooms on the ground representing the periphery of mycelial growth of a basidiomycete.

- **falling drop technique**
  a technique that provides a measure of the stability of structural units to raindrop impact.

- **fallout, dust fall, particle fall**
  dust that falls from the sky, often in rain or snow. Whether coarse or fine, it may originate from chimneys, volcanoes, sandstorms, forest fires or nuclear explosions.

- **fallow**
  temporarily relieving land from crop cultivation for some time to enable it to recoup soil fertility and ensure heavy yields of the succeeding crops. The period of relief may vary from a crop season to a year. The soil is usually cultivated to control weeds and conserve moisture.

- **fallow land**
  cultivable land not cultivated.

- **family labour**
  labour contributed by the members of the farmer's family.

- **famine**
  extreme scarcity of something, especially food, existing on a group scale.

- **fancy sugar**
  refined sugar pressed into cube or tablet form. Fine grained refined sugar for quick dissolving. Icing sugar, ground and sifted refined crystals. Coffee crystals, large refined crystals coloured and flavoured with caramel. Sugar candy— very large crystals grown slowly, refined syrup.

- **fanning mill**
  a hand or electrically operated machine equipped with a hopper, vibrating sieves, a fan etc., for removing chaff, straw, weed seeds and dirt from grain.

- **far infrared**
  the part of the spectrum towards the end of the infrared region of the electromagnetic spectrum which is near the radio wave region.

- **far red light**
  light of wavelength about 730 nm which reverses the effect of red light on phytochrome.

**farm**
a tract or tracts of land operated under one management as a single working unit for carrying on one or more agricultural enterprises such as the growing of crops or the raising of domestic animals.

**farm building**
any service building on a farm such as tool shed, store, poultry house, etc.

**farm bureau**
an organisation of farmers and those interested in agriculture for the improvement of agriculture and rural life.

**farm business**
the series of production and maintenance activities and financial transactions in connection with running a farm.

**farm capital**
the average value of all real estate, implements, machinery, livestock, cash and other investment taken at the beginning and ending of the farm year.

**farm crops**
economic crops grown on a field scale on a farm as different from truck crops, fruit crops, etc.

**farm duty of water**
the seasonal quantity of water delivered to individual farm units under an irrigation project. Such duty may be expressed in terms of acre-feet per acre, total depth of water over a unit area, or number of acres served per unit of flow. Such duty expresses the actual rate of use of water at the farm, after all canal and other losses have been eliminated, but includes losses in the farm ditches and also waste.

**farm economy**
the practical adjustment, organisation, administration of the affairs and business of a farm.

**farm family**
a farm of sufficient size to provide employment for its operator and the members of his family, who themselves

perform the entrepreneurial and managerial functions and also do all, or at least the bulk, of the manual labour on the farm.

- **farm forestry**
  it includes
  1. creation of wind breaks around the farms to conserve moisture, to meet fuel requirements, fodder etc., raising of shelter belts on large Government lands across the wind diversion because of small holdings of individuals.
  2. creation of farm woodlots in certain percentage of the farmers land for fuel, fodder and small timber.

- **farm harvest price**
  the average wholesale price received by the farmer for the commodity produced by him at the village site during the specified harvest period.

- **farm income**
  income obtained from farm operations. The sum of receipts from sales, appreciation of inventory and value of living obtained from the farm represent gross farm income. After expenses are deducted, the reminder is called net farm income.

- **farm irrigation requirement**
  it depends on the irrigation need of individual crop, their area and the losses in farm water distribution systems, mainly through seepage.

- **farm labour**
  a person who is gainfully employed by a farm operator to assist with the farm work. He may be a wageworker, a family member, or other person working for perquisites. He may be regular or seasonal, local or migratory, full time or part time.

- **farm losses**
  losses of water on the farm due to uneven distribution, poor handling, evaporation and percolation into the subsoil due to over-irrigation.

- **farm management**
  1. the science which considers the organisation and operation of the farm from the point of view of efficiency and continuous profit.
  2. utilisation of sound principles in the selection, organisation and conduct of an individual farm business for the purpose of obtaining the greatest possible profit.

- **farm manager**
  a man in charge of an agricultural enterprise large enough to necessitate use of his full time in supervision and business details who selects the crops and products, organising of work, its proper execution, the care of labourers and selling produce, keeping farm records, financial transactions etc.

- **farm map**
  the map of a farm that gives the clear picture of the situation of blocks, plot or subplots, farm buildings, wells, underground pipe lines, drainage, roads and bunds, fencing etc. which facilitate in planning the cropping scheme.

- **farm operator**
  a person who manages and directs the operation of a farm.

- **farm peaks**
  the amount of crop a farmer is permitted to grow. They may be expressed in terms of tons of yield or area to be harvested.

- **farm planning**
  a process involving many decisions to be taken in respect of the kinds of crops to grow, rotations, mixtures, soil and water conservation practices to be followed, and buildings, bullocks machinery purchase etc.

- **farm pond**
  are the bodies of water made either by constructing a dam or an embankment across a water course or by excavating a pit or dug out.

- **farm population**
  all persons living on farms regardless of occupation.

- **farm processing**
  any method or treatment used to prepare farm products for use or preservation is called farm processing. It includes the operation that maintains or improves the quality of farm products or changes their forms. It includes the operations like chart cutting, grain grinding, cane crushing, grain drying, milk processing, seed processing etc.

- **farm seed**
  seed from a previous crop retained for own use and not necessarily conforming to the seeds Act.

- **farm waterway**
  a natural or constructed waterway which is entirely within an individual farm boundary.

- **farm yard manure (FYM)**
a mixture of one or more organic matter like cattle dung and urine, litter, remnants of straw, plant stalks and leaves fed to the cattle.

- **farmer**
a person who cultivates land for business.

- **farmer's earning**
the amount of labour income plus what the farm furnishes towards the living of the farmer and his family.

- **farming system**
1. the entire complex of resource preparations, allocations, decisions and activities which within an operational farm unit or a combination of such units, results in agricultural production.
2. the harvesting, drying, processing and marketing of the products are also directly related to the system that produces them.

- **farmstead**
that part of the farm which is occupied by the main set of buildings, adjacent yards, kitchen garden and family orchard.

- **farmyard**
a small area which is adjacent to a cattle stall and in which animals are kept when not in pasture, at work or in the stall.

- **fast neutrons**
neutrons having energies in excess of about 0.1 MeV.

- **fat clay**
clay of relatively high plasticity. Opposite: lean clay.

- **fats**
are esters produced by the condensation of three molecules of it or of different fatty acids with glycerol. Plant fats are liquids at ordinary temperatures, and are, therefore, called fatty oils, while the solid fats are referred to as fats proper. Only those fatty acids which contain an even number of carbon atoms (*e.g.* acetic, butyric, caproic, palmitic, stearic, oleic, linolenic and ricinoleic) occur in fats. Fats are present in all living cells, but are most abundantly found in the storage tissues of oily seeds. These fats are produced from sugars translocated into the seeds during ripening. During germination of the seed the reverse process takes

place. All natural fats are mixtures of glycerides.

- **fatty acid oxidation**
  a process occurring in the mitochondria of higher animals and in plants that store fat, consisting of the oxidation of fats resulting in the production of ATP.

- **fatty acids**
  a group of saturated and unsaturated monobasic aliphatic carboxylic acids which form esters with glycerol and other alcohols to make fats, oils, waxes, and other lipids.

- **faurial pedoturbation**
  is soil mixing by animals such as .ants, earthworms, moles, rodents, and man himself.

- **fe horizon**
  a horizon enriched in iron oxide and hydroxide.

- **feather analysis**
  1. a technique first described by N.
  2. Feather in 1938 for measuring the range of beta particles by comparison with beta particles of known energy distribution.

- **feeder roots**
  fine roots with a large absorbing area (root hair).

- **fehling's solution**
  a solution used to detect the presence of aldehydic group (-CHO). It is a mixture of freshly prepared copper (11) sulphate solution with alkaline potassium sodium 2,3-dihydroxy butanedioate (tartrate). The aldehydic group gives red precipitate on heating with Fehling solution due to the : formation of cuprous oxide.

- **felsapthoids**
  the minerals lucite (K2$Al_2Si_4$Oi2), nephelite (Na-K silicate) and sodalite (3 Na. $Al_2SiO_8$. 2NaCl) constitute a felspathoid group of minerals bearing close relationship and unproductive soils.

- **fen**
  a low-lying land area partly covered by water.

- **fencing**
  fixing the boundary of the farm to stop encroachment of land from wild animals and cattle.

- **ferment**
  a material produced by a living organism which catalyses certain reactions and which consists of an enzyme in association with other substances of the organism.

- **fermentation**
  decomposition occurring in organic compounds, usually carbohydrates, brought about by enzymes either directly or in micro-organisms, resulting in the production of other compounds such as alcohol or lactic acid, often with heat and gases, often used for anaerobic spiration, especially of yeast, with the production of alcohol.

- **fermentation layer**
  in soils, a layer of partly decomposed litter lying below the freshly fallen organic material and above plant remains that have been fully decomposed to form humus. This stratified structure is typical of mor humus.

- **ferrallite**
  well-drained soils formed from basic igneous rocks, consisting mainly of hydrated oxides of iron and aluminium, with low silica content.

- **ferredoxin**
  a group of red-brown proteins of green plants, blue-green algae, and some bacteria. They contain nonhaem iron in association with sulphur at the active site. They are strong reducing agents and function as electron carriers as in photosynthesis and nitrogen fixation. They are common components of electron transport chains in most organisms.

- **ferrihumic**
  it consists of authigenic deposits (formed in place) of hydrated iron oxides mixed with varying kinds or amounts of organic materials.

- **ferrisols**
  a group of red-coloured soils occurring in the humid tropics that resemble ferrallitic soils in structure and development but are less intensely leached and weathered. They are formed where surface erosion is constantly removing

weathered material from the surface (*e.g.* on slopes). They are more fertile than ferrallitic soils.

- **ferritisation**
formation of a horizon of accumulation of iron oxides by migration of iron in soils.

- **ferro**
prefix denoting iron, especially in names of alloys, *e.g.* ferro-manganese.

- **ferrugination**
release of iron from primary minerals and the dispersion of particles of iron oxide in increasing amounts, their progressive oxidation or hydration, giving the soil mass brownish, reddish brown, and red colours, respectively.

- **ferruginous soil**
soil containing large amounts of iron minerals, especially linonite and hematite. Characterised by red colour or shades of yellow and brown.

- **fertigation**
application of fertilisers with irrigation water. Its applicability depends on external conditions, such as the type of soil and crop, method of irrigation used, water quality, types of fertiliser available, and general economics when compared to other ways of fertiliser application.

- **fertilisation (plant).**
1. the union of the male (pollen) nucleus with the female (egg) cell.
2. the process of adding plant food to the soil for purposes of plant growth.

- **fertiliser**
any organic or inorganic material of natural or synthetic origin that is added to a soil to supply elements essential to plant growth. The three major elements applied through fertilisers are nitrogen, phosphorus and potassium. They are intended to supply plant needs directly rather than indirectly through modification of such properties as pH.

- **fertiliser analysis**
as applied to fertilisers, designates the actual percentage composition of the product as determined by a laboratory analysis.

- **fertiliser application**
the use of materials containing plant nutrients on soil.

- **fertiliser application rates**
determining the optimum amount of fertiliser requires

experience that can serve as a basis for predicting how the crop will respond to fertiliser applied to the soil on which it is grown. The yield increase per unit of additional fertiliser declines at the higher rates of fertilisation. A point is reached where the additional fertiliser barely increases production enough to be worth its cost. This point is the maximum fertiliser rate that should be considered. A farmer may rationally choose either the minimum or the maximum fertiliser rate or possibly some rate between these limits. The maximum rate gives the largest profit per unit of land and is the best rate if enough capital is available. The minimum rate gives the highest return per rupee of fertiliser investment and is the best choice if money for fertiliser is limited. A higher fertiliser price gives a steeper cost line and reduces the size of the profit section on the graph.

- **fertiliser attachment**
device attached to an implement or machinery to facilitate distribution of fertiliser in or upon the soil.

- **fertiliser brand**
a term, design, or trademark used in connection with one or several grades of fertilisers.

- **fertiliser brand name**
a specific designation applied to an individual fertiliser.

- **fertiliser conditioners**
materials which are added to fertilisers to control free moisture and keep the fertiliser free flowing by helping to prevent caking.

- **fertiliser formula**
the quantity and grade of materials used in making a fertiliser mixture.

- **fertiliser formulations**
a fertiliser formulation is the act of or result of mixing specific fertiliser materials of known physical and chemical composition to produce a predictable fertiliser containing two or more plant nutrients.

- **fertiliser grade**
the minimum guarantee of plant food content of a fertiliser mixture or material expressed in terms of total N, available $P_2O_5$ and water soluble $K_2O$.

- **fertiliser ratio**
fertiliser analysis are sometimes reduced to fertiliser ratios for the purpose of comparing relative amounts of each nutrient supplied by the

fertiliser. It is obtained by dividing the analysis by a factor that produces the smallest possible whole numbers. A 3-12-12 fertiliser thereby becomes a 1-4-4 ratio. A 5-20-20 fertiliser also has a 1-4-4 ratio. These two fertilisers could therefore be substituted for one another by adjusting the rate of application.

- **fertiliser requirement**
the quantity of certain plant nutrient elements needed, in addition to the amount supplied by the soil, to increase plant growth to a designated optimum.

- **fertiliser use efficiency**
a measure of the extent to which fertilisers and management practices have worked together with the soil-climate complex in producing a high yield. It is closely linked with the efficiency of the total yield production system. Thus, fertiliser use efficiency rests heavily on the removal of gains from positive interactions.

- **fertility (of soil)**
the quality that enables a soil to provide the proper compounds, in the proper amounts and in the proper balance, for the growth of specified plants when other factors, such as light, temperature, and the physical conditions of the soil, are favourable.

- **fertility gradient**
the variation in natural fertility in any direction across an area of ground.

- **fibre crops**
the fibre crops include cotton, flax, hemp, ramie, kenaf, phormium and sunnhemp.

- **fibre fineness (cotton)**
weight per unit length of the fibre.

- **fibre maturity (cotton)**
indicates the degree of thickening of the cell wall in relation to its diameter. Fibres having thick cell walls and comparatively a narrow lumen are termed mature, while those having practically no secondary thickening or extremely thin walls and a wide lumen are called immature fibres, those lying intermediate between these two classes, are called half mature fibres.

- **fibric soil material (diagnostic organic material formerly peat)**
in an unrubbed condition, fibres compose over 2/3 of the mass, and the material yields almost clear solutions when

extracted with sodium pyrophosphate.

- **fibrous root**
a fibrous root system consists of a tuft of adventitious roots of more or less equal diameter, arising from stem base or hypocotyls, and bearing smaller lateral roots *e.g.* wheat, strawberry.

- **fid fidus**
originally a Latin word that means a 'cleft'.

- **fiducial limits**
in fiducial inference limits between which a parameter is considered to lie. Theoretically, different but in most practical cases the same.

- **field capacity**
the moisture percentage, on a dry weight basis, of a soil after rapid drainage has taken place following an application of water, provided there is no water table within capillary reach of the root zone. This moisture percentage usually is reached within one to three days after an irrigation, the time interval depending on the physical characteristics of the soil.

- **field capacity zone**
a buffer zone in which water is held against gravity under free drainage. It is this zone which prevents the moisture and consequently the salts from moving in a vertically upward direction.

- **field crops**
herbaceous plants grown in cultivated fields under more or less extensive system of culture.

- **field duty of water**
the quantity of water delivered to individual fields or orchards under an irrigation project. It is expressed in terms similar to those for farm duty of water.

- **field efficiency (plough)**
it is the ratio of effective field capacity and theoretical field capacity, expressed in percentage.

- **field germination**
the percentage, by number, of seeds in a given sample that germinate and appear above the soil surface in the field or nursery, irrespective of later survival.

- **field irrigation efficiency**
is the percentage of the water delivered at the edge of the field that is actually added to the root zone and is available for evapotranspiration.

▪ **field irrigation requirement of a crop**
the water requirement of crops exclusive of rainfall and contribution from soil profile.

▪ **field moisture**
the per cent moisture content in a field sample of soil at any one time.

▪ **field moisture capacity**
the approximate quantity of water which can be permanently retained in the soil in opposition to the downward pull of gravity. It may be expressed in per cent of dry weight or in inches depth for a given depth of soil. The length of time required for a soil to reach field moisture capacity varies considerably with various soils, being approximately 24 to 48 hours for sandy soils, 5 to 10 days for silt clay soils, and longer for clays. Also called capillary capacity, field carrying capacity, maximum water holding capacity, moisture holding capacity, and normal moisture capacity.

▪ **field moisture deficiency**
the quantity of water which would be required to restore the soil moisture content to field moisture capacity.

▪ **field moisture equivalent**
the minimum moisture content, expressed as a percentage of the weight of the oven-dried soil, at which a drop of water placed on a smoothed surface of the soil will not immediately be absorbed by the soil but will spread out over the surface and give it a shiny appearance.

▪ **field permeability coefficient**
the rate of flow of water, in gallons a day, under prevailing conditions, through each foot of thickness of a given aquifer in a width of 1 mile, for each foot per mile of hydraulic gradient.

▪ **field strip cropping**
the practice of growing crops in strips across the general slope but parallel to a field boundary rather than on the contour. The strips are uniform in width and easy to cultivate and harvest, but the effectiveness of control of water erosion is reduced by the deviation from the contour.

▪ **field system**
the system of farming practised in Britain and parts of Europe during the Middle Ages. Arable land was divided

into strips and members of the community cultivated strips in different parts of the village. Animals were grazed on an area of common land. When the land was cultivated one year and left fallow the next it is called the two-field system, cultivation for two years and fallow for one is called the three-field system.

- **field test**
  an experiment conducted under field conditions. Ordinarily less subject to control than a formal experiment and may be less precise.
- **field water balance**
  the water balance of a field is an itemised statement of all gains, losses, and changes of storage of water occurring in a given field within specified boundaries during a specified period of time.
- **field water losses**
  is the water lost through evaporation, deep percolation, surface runoff and wind drift. These losses, to some extent, can be controlled by proper design and operation of an irrigation system.
- **field water use efficiency (EU)**
  the ratio of crop yield (Y) to the total amount of water used in the field (WR). It is expressed as follows: Crop yield can be defined either in terms of growth *i.e.* dry matter produced or in terms of marketable product.
- **fifteen atmosphere value**
  the percentage of water remaining in the soil when it is subjected to a pressure of 15 atmosphere.
- **fifty per cent yield decrement value**
  the measured value of the soil salinity or alkali that decreases crop yield 50 per cent as compared with yields of the same crop on non-saline and non-alkali soil under similar growing conditions.
- **filler**
  is a make-weight material added to a mixed fertiliser or fertiliser material to make up the difference between the weight of the added ingredients required to supply the plant nutrients in a ton of a given analysis and 1000 kg.
- **film**
  a thin layer of adsorbed water on the surface of the grains of the soil.
- **film badge**
  a piece of photographic film

contained in a lightproof holder and worn by an individual in order to measure the amount of radiation to which he is exposed.

- **film water**
  the water held on the surface of soil particles that does not drain away, although it moves rapidly under suction gradients. Most of it is available to plant roots.
- **filter crop**
  close-growing crops planted across the slope to retard runoff and cause deposition of the soil load.
- **filter point wells**
  wells generally drilled in delta and coastal areas where the substrata is soft and sandy. For constructing these wells bores are drilled up to a depth of about 30 ft. from the ground level and galvanised iron tubes are driven into these bores. As the first tube driven in the bore has a perforated end known as 'filter point', these wells are commonly called as 'filter point wells'.
- **filter points**
  in deltaic regions, where the aquifer formations are of coarse sand and gravel, the tube wells are shallow and consist of a well screen and a short length of casing pipe. These wells are called filter point wells.
- **filter strip**
  a strip of permanent vegetation of sufficient width and vegetative density above farm ponds, diversion terraces and other structures to retard flow of runoff water, causing it to deposit soil, thereby preventing silting of the structure or reservoir below.

- **filtration**
  in waste water treatment, the mechanical process that removes particulate matter by separating water from solid material usually by passing it through sand or any other filter.
- **final crop**
  that portion of the stand which is intended to retain to the end of the rotation.
- **final host**
  the host in which a parasite develops to maturity. final

intake of water amount of water which enters the soil after a constant rate of intake has been developed. Varies with kind of soil. Measured by ring infiltometer.

- **final yield**
  all the material that counts against the prescribed yield derived from the principal fellings in a regular forest.

- **fine earth**
  refers to particles smaller than 2 mm in diameter.

- **fine texture**
  consisting of or containing large quantities of the fine soil fractions, particularly silt and clay. Includes sandy clay, silty clay, and clay textural classes.

- **finite tolerance**
  the maximum amount of pesticide which can legally remain on a food or feed crop at harvest after the pesticide has been directly applied to the crop.

- **fire clay**
  a clay which is capable of withstanding high heat, and is used for brick making.

- **fire climax**
  a plant community maintained as climax vegetation by natural or man-made fires which destroy the plants that would otherwise become dominant.

- **fire forest**
  any fire not employed as an indirect means of forest protection or management and which occurs on forest land.

- **fire ground (forestry)**
  a fire that not only consumes all the organic materials of the forest floor, but also burns into the underlying soil itself, as, for example, a peat fire (Usually combined with, but not to be confused with, a surface fire).

- **fire harrowing**
  burning of organic material by dragging harrows containing lighted accumulated material over the ground surface used where there has been a poor initial burn.

- **fire line**
  a cleared permanent fire break intended to prevent fires crossing from one area into another.

- **fire protection**
  all activities concerned with protection of a forest area from damage by fire.

■ **fire season**

1. the period of the year during which forest fires are likely to occur and become dangerous.
2. the period or periods of the year during which the use of fire in a forest is subject to legal restrictions.

■ **firm**

a consistence term used in describing most soil which crushes under moderate pressure between thumb and forefinger, but with a resistance distinctly noticeable.

■ **first bottom**

the normal flood plain of a stream.

■ **first order correlation coefficient**

any correlation coefficient based on the relationship of three variables. It may be either a partial or a multiple correlation coefficient in which only three variables are involved.

■ **first order interaction**

the interaction between two variables or two main effects. Also called simple interaction.

■ **first order reaction**

a reaction in which the rate of reaction is proportional to the concentration of one of the reacting substances for example hydrolysis of ester.

■ **first wilting percentage**

the water content of the soil at the time of wilting of basal leaves and cessation of elongation of growth.

■ **fish manure**

a manure prepared from non-edible fish, carcasses of fish and offal. Fish-boiler and squeezer are used for preparing fishmeal. The process consists in drying, powdering and filling the bags. Fish meal analysis 5-10%N, 4-9% $PaO_5$ and 0.3 to 2.0% $K_2O$. The manurial composition of fish meal varies with the type of fish employed. The C/N ratio of fish meal is 4 to 5, and as such it readily decomposes in soil.

■ **fish silage**

fish converted into winter feed for livestock through the process of fermentation into airtight chamber where it is compressed to exclude air and where it undergoes an acid fermentation that retards spoiling.

■ **fish tankage, fish scrap, dry ground fish, fish meal**

the dried, ground products derived from rendered or unrendered fish.

- **fishing**
the selection and removal of a microbial colony for pure culture study.

- **fission**
the splitting of a heavy nucleus into two approximately equal pieces.

- **fixation**
the process or processes in a soil by which certain chemical elements essential for plant growth are converted from a soluble or exchangeable form to a much less soluble or to a nonexchangeable form.

- **fixation of nitrogen**
the formation of nitrogenous compounds from gaseous nitrogen, by soil bacteria.

- **fixation of phosphorus**
the conversion of soluble phosphorus fertilisers in the soil into less soluble and available forms.

- **fixatives**
absolute methyl alcohol or glacial acetic acid may be used in place of a flame to fix slides before staining.

- **fixed ammonium**
ammonium held in a soil which is neither water-soluble nor easily exchangeable.

- **fixed copper**
are formulated at the same content as boardeaux. In all these, copper is stabilised and rendered less soluble to reduce the phytotoxicity of copper.

- **fixed cost**
expenses which have to be incurred every year in a farm irrespective of whether the farming operations are carried out or not *e.g.* revenue, taxes, interest on borrowed capital and rent.

- **fixed ground water**
ground water held in saturated material with interstices so small that it is permanently attached to the poor walls, or moves so slowly that it is usually not available as a source of water for pumping.

- **fixed input**
The term refers to that factor of production which is used by the firm in a fixed quantity in the short run. Examples are—land, building, plant etc. In the long run there are no fixed inputs because the firm can vary its scale of operations as well as its output.

- **fixed input coefficients**
the situation where the production of any level of output requires a fixed proportion of factor input.

■ **fixed light position**
the final position of a fully developed leaf in relation to the direction of light falling on it.

■ **fixed moisture**
moisture held in the soil below the hygroscopic limit.

■ **fixed phosphorus**
1. that phosphorus which has changed to less soluble forms as a result of reaction with soil, moderately available phosphorus.
2. applied phosphorus, phosphorus not taken up by plants during the first cropping year.
3. soluble phosphorus, phosphorus attached to the solid phase of soil in forms highly unavailable to crops.
4. unavailable phosphorus, phosphorus other than readily or moderately available forms.

■ **fixed potassium**
potassium held by a soil which is neither water-soluble nor readily exchangeable.

■ **fixing**
the procedure of making microorganisms adhere to the surface of a glass slide by allowing a smear to dry and then heating slightly over a flame.

■ **flag leaf**
the last fully expanded leaf in cereals. Before the head emerges the flag leaf surrounds the head and is then called a boot leaf.

■ **flag stage**
the early post-emergence stage of onion seedlings between the crook stage and the emergence of the first true leaf. The bent tip of the seed leaf resembles a flat attached to a staff. Also used to designate the stage of development in cereals and other grasses at which the sheath and leaf have been produced.

■ **flaggy**
soils contain relatively thin fragments 6 to 15 inches long of sandstone, limestone, slate, or shale, or, rarely, of schist. A single piece is a flagstone.

■ **flagstone**
a relatively thin fragment 6 to 15 inches long, of sandstone, limestone, slate, shale, or schist.

■ **flat rate**
the charge for water that is not metered. It may be based on size of service pipe, number of water fixtures or outlets, land frontage, or other arbitrary measures.

- **flavonoids**
any of various compounds containing a $C_6$-$C_3$-$C_6$ skeleton, the $C_6$ parts being benzene rings and the $C_3$ part varying in different compounds, and including many water-soluble plant pigments.

- **flavoproteins**
proteins with a firmly attached flavin prosthetic group, important in electron transfers such as respiratory oxidations, as they can be alternately reduced and oxidised, and which are yellow when oxidised but colourless when reduced and are also called the yellow, or flavin, enzymes. Flavoproteins are enzymes of the dehydrogenase type.

- **flexibility of prices**
under perfect competition, there is flexibility of prices. Any changes in the conditions of demand and supply of the commodity, lead to changes in it's price. But under modern conditions of imperfect competition, there is a tendency for prices to become less flexible.

- **flexible conduit**
a conduit whose cross-sectional shape can be distorted sufficiently to change its vertical or horizontal dimension more than 3 per cent before causing materially injurious cracks, including those made of corrugated pipe, thin steel, or wrought iron pipe.

- **flexible joint**
any joint between two pipes that permits one of them to be deflected without disturbing the other pipe.

- **float gaging**
the process of measuring the velocity of water flowing in a stream or conduit by means of floats placed in the water.

- **float valve**
a valve wherein the plug is actuated by a float to control the flow into a tank.

- **flocculated soil**
a soil in which the solum is more or less acid, but which shows no pronounced migration of clay.

- **flocculating agent**
a reagent added to a dispersion of solids in a liquid to bring together the fine particles to form floes.

- **flocculation**
the grouping of solids in water, resulting in floes, often wrongly thought to be the same as coagulation. It can be promoted by gentle stirring (mechanical flocculation) or

by adding chemicals (flocculants).

- **flocculation agent**
a coagulating substance which when added to water, form a flocculent precipitate which will entrain suspended matter and expedite sedimentation (For example, Alum, ferrous sulphate, lime).

- **flocculation limit**
the water content of a soil, when it is in the condition of a deflocculated sediment.

- **flocculation ratio**
the void ratio of a soil when it is in the condition of a deflocculated sediment.

- **flocculent**
an adherent flaky or fluffy aggregate of any materials (including micro organisms floating in or on a liquid).

- **flood**
an overflow or inundation that comes from a river or other water body mainly as a result of heavy rainfall or snow-melt and causes or is likely to cause damage to crops, life and property.

- **flood coefficient**
a numerical expression of the proportion of the pluviograph runoff that actually occurs as surface runoff, expressed as a decimal, the per cent runoff obtained by dividing the pluviograph runoff by the actual surface runoff.

- **flood irrigation**
1. a flooding of the land surface in irrigation.
2. a system of irrigation utilising the natural flood rises to spread water over the land by means of canals and ditches, found particularly in older irrigated areas of the world.

- **flood plain**
in a valley bottom, an expanse of flat land which is usually dry, but submerged during heavy flood. Flood deposits are usually fertile silts from upstream erosion.

- **flood stage**
the state of river flow beyond bank full stage.

- **flooding injury**
moisture may injure or kill plants if present in excess, but this is actually due to lack of

oxygen for root respiration and is called flooding injury. This results in, the 'physiological drought' in plants.

- **flooding soil**
  1. a condition in which there is free water present at the surface of the soil.
  2. water from field channels is allowed to flood the fields in irrigation by flooding.

- **flora**
  the plant life of any region.

- **floral pedoturbation**
  is soil mixing by plants as in tree tipping that forms pits and mounts.

- **florigen**
  a hypothetical, or as yet not isolated, phytohormone which causes a bud to develop into a flower, postulated as the stimulus for flowering that is perceived in the leaves and must be transmitted to the tip, and which may be a mixture of other hormones.

- **flos**
  a Latin word that means 'flower'.

- **flotation sulphur**
  is a form of sulphur fungicide obtained as a by-product of the manufacture of fuel gas from coal. It is recovered by the flotation process.

- **flow**
  1. the movement of a stream of water or other mobile substance from place to place, a stream of water, movement of silt, water, sand etc.
  2. the fluid which is in movement.
  3. the quantity or rate of movement of a fluid, discharge, total quantity carried by a stream
  4. to issue forth or discharge.

- **flow diagram (flow chart)**
  a diagram that shows a sequence of interlinked operations, events, etc.

- **flow meter**
  in waste water treatment, a meter that indicates the rate at which waste water flows through the plant.

- **flow production**
  it is an alternative term for Mass Production.

- **flow rate**
  the rate at which a fluid flows, or is discharged, usually measured in cubic feet per second, gallons per minute, etc. Same as discharge rate.

- **flow velocity (of water in soil)**
  the volume of water transferred per unit of time and per unit of area normal to the di-

rection of the net flow.

- **flowability**
  of the dust is an indicator of the feed rate of dusting equipment. The angle of slope is a measure of flow-ability.
- **flowering hormones**
  are hormones that initiate the formation of floral primordial or promote their development.
- **flowering regulators**
  are regulators that affect flowering.
- **flowering stage**
  the stage of development of a plant when it begins to produce flowers.
- **fluctuations cyclical**
  these are ups and downs of business-activity, employment and prices.
- **flue**
  the special passageways for heated air in a barn for curing tobacco.
- **flue curing (tobacco)**
  a curing process divided into four stages
  1. yellowing or colouring—this is the accelerated ripening stage, temperature and moisture conditions must be favourable to growth and should therefore resemble hot-house conditions.
  2. fixing or killing—the dry tips indicate that the leaf is commencing to dry, when this stage has been reached the fixation of colour can commence.
  3. drying out—at this stage the midrib or stem still contains moisture and to complete the cure this must be removed.
  4. conditioning—before the leaf is removed from the barn it must be brought to such a stage that it can be handled without damage. Conditioning is right when the web of the leaf and the midrib half way back from the tip are pliable and the reminder of the midrib is slightly supple but will break if bent at right angles.
- **flue dust**
  is a by-product from steel and cement industry and is used as a source of potassium.
- **fluid formulations (fertilisers)**
  may be solutions (clear liquids) or suspensions (salt suspensions, colloidal suspensions, or thin slurries), aqua ammonia (anhydrous ammonia plus water), nitrogen solutions (dissolved solids or solids plus ammonia) and so-

lutions of dissolved mixed fertilisers.

- **fluidity of labour**
  it is an alternative term for 'mobility of labour'.

- **fluitans**
  a Latin word that means 'floating'.

- **flumes**
  are used to carry irrigation water across streams, canals gullies, ravines or other natural depressions. They may be made of open channels or pipes which are often supported by pillars or may be affixed to bridges.

- **fluorescence**
  the extranuclear emission of photons caused by excitation of an atom.

- **flushed soil**
  is a type of soil on lower hillslopes which are rich in bases due to downhill movement of products of leaching from hillsides.

- **flux**
  the amount of an entity received by or transferred across a particular plane per unit time.

- **flux density**
  the amount of an entity received by or transferred across a particular plane per unit area per unit time.

- **foaming agent**
  a material which causes a pesticides mixture to form a thick foam. It is used to reduce drift.

- **fodder**
  coarse grasses such as corn and sorghum harvested with the seed and leaves and cured for animal feeding.

- **fodder crop**
  a crop which is purposely grown for feeding animals.

- **fog precipitation**
  the precipitation of liquid water from non-rain bearing clouds due to interception of cloud particles by trees and/or other vegetation.

- **foliage**
  also called leafage. The leaves, collectively, of a plant.

- **foliage plant**
  plant grown for the colour and/or shape of its foliage.

**foliar diagnosis**
an estimation of the extent to which plants are getting certain necessary chemical elements from the soil by examination of the colour and growth habits of the foliage of the plants.

**foliar feeding**
applying liquid nutrients to the leaves.

**food chain**
sequence of organisms in which each is food of a later member of the sequence.

**food corporation of india**
a statutory body set up by the government of India in 1965 to undertake purchase, storage, movement, transport, distribution and sale of foodgrains and other foodstuffs.

**forecast**
a statement of anticipated meteorological conditions for a specific place (or area) and period of time.

**forest**
1. the word 'Forest' is derived from the Latin word 'Foris' which means 'outside'. Therefore, forests are areas covering practically all uncultivated or untended lands covered with rather tall and dense tree growth. The word 'jungle' denotes whole range of trees and shrub growth with a suggestion of wilderness.
2. an area set aside for the production of timber and other forest produce and is under woody vegetation conferring certain benefits on the society.
3. legally, forest areas which have been notified by the government, as reserved forest(s) and they may include some blank patches.
4. ecologically, the term forest signifies a complex organism, composed of distinct biological units called forest communities, that have come into being by the combined action, reaction and coactions of a variety of organisms with the complex factors of the habitat that themselves change, both in space and time.

**forest communal**
a forest owned and generally managed by a community such as a village, town, tribal authority or local Government, the members of which share in the produce or proceeds.

**forest floor**
all dead vegetable matter on the mineral soil surface. Includes

litter and unincorporated humus.

- **forest soil**
great variation, fine in texture surface layer consists of well decomposed humus and mineral soils, strongly acidic in reaction, high base exchange capacity is due to high organic matter content, water soluble Mn is present, suitable for forest trees.

- **forest virgin**
natural forest essentially uninfluenced by human activity, it may or may not be a climax forest.

- **forest, secondary**
forest growth that has come up naturally after partial or complete removal of a crop by cutting, fire or other causes. It is frequently different in composition and character from the growth it replaces.

- **forestalling**
it means cornering the market in a commodity in order to resell it at a much higher price.

- **forestry**
the theory and practice of constitution and management of forests and utilisation of their products. It also stands for scientific management of forests for the continuous production of goods and services.

- **furrow grade**
slope or grade of furrow for irrigation, controls speed at which water flows down the furrow, minimum of 0.05 per cent needed to assure surface drainage, under other conditions, 0.2 to 0.5 per cent.

- **furrow irrigation**
a method of applying irrigation water to fields or orchards by small ditches or furrows which lead from the supply ditch.

- **furrow slice**
the mass of soil cut, lifted and thrown to one side of an implement.

- **furrow sole**
the bottom of the furrow on which the plough bottom slides.

- **fused tricalcium phosphate**
a product resulting from the fusion of phosphate rock to produce the alpha form of tricalcium phosphate.

- **fusion**
the joining of light nuclei to form a heavier nucleus. Particle velocities corresponding to millions of degrees are required for this process.

**fuzz**
cotton fibre, 1/64 to 1/16 inch long, thicker than lint and has no twists, does not get separated from cotton seed at the time of ginning, not much use commercially.

**gel electrophoresis**
electrophoresis performed in silica gel, a porous inert medium.

**gene**
1. the functional and physical unit of heredity.
2. A gene is a sequence of nucleotide bases located in a particular position on a particular chromosome that encodes for a specific RNA molecule or protein. Each represents a unit character, which is recognised by its effect on the individual bearing the gene in its cells. There are many thousand genes in each nucleus.

**gene centres**
geographical regions in which certain species of cultivated plants are represented in the greatest possible diversity of forms.

**gene complex**
1. the balanced system of genes constituting the internal medium within which each gene can manifest its effect.
2. a group of genes which combine to determine the development of a particular character.

**general purpose**
pertaining to an animal, tool, implement, etc. being of use for or put to several purposes.

**genetic thread principle**
the theories of soil genesis provide a framework for aiding in determining the significance and relevance of soil properties for use as differentiating characteristics. This principle is followed in the use of theories of evolution as a framework for taxonomy in the plant and animal kingdoms.

**genotype**
1. the genetic makeup of an organism, as characterised by its physical appearance or phenotype.
2. The genetic constitution of an organism or a species in contrast to its observable characteristics (the latter is the phenotype).

**geobotany**
branch of botany dealing with all aspects of relations between plants and the earth's surface, comprises plant ecology and

plant geography.

- **geodesy**
the science of determining the size and shape of the earth and the precise location of points on its surface.

- **geohydrology**
that branch of hydrology relating to subsurface of subterranean waters.

- **geologic erosion**
is a normal process representing erosion of land in its natural environment without the influence of man. It is chiefly caused by the action of wind, water, temperature variation and glaciers.

- **geology**
the study of the planet earth — the materials it is made of, the processes that act on those materials, the products formed, and the history of the planet and its life forms since its origin.

- **geomagnetism**
is the study of earth's magnetic field.

- **geometric similarity**
similarity of form which exists between two objects or systems if the ratio of all corresponding linear dimensions are equal. This relationship is independent of motion of any kind.

- **geometry factor**
the solid angle at the radioactive source subtended by the sensitive volume of the detector divided by the total solid angle.

- **geotome**
an instrument for taking soil-samples, without disturbing the surrounding soil.

- **germ cell**
a cell capable of reproduction, or of sharing in reproduction.

- **germ plasm**
a kind of protoplasm which, according to Weismann, was transmitted unchanged from generation to generation in the germ cells.

- **germicide**
a substance formulated to kill germs or microorganisms. The germicidal properties of chlorine make it an effective disinfectant.

- **germination cabinet**
a type of seed germinator most commonly used in the seed testing laboratories.

- **germination percentage**
a general term for the percentage of seeds in a given sample that actually germinate.

- **germination test**
a statistical test in which the germinative power of seeds and the effects on germination of various treatments applied to the seeds, are examined.

- **germinative capacity**
the percentage, by number of seeds in a given sample that actually germinate, irrespective of time or that germinating within twice the energy period. The total number of seeds that germinate in a germinator, plus the number of seeds remaining at the end of the tests, expressed as a percentage of the total sample.

- **germinative power**
the percentage, by number, of seeds in a given sample that actually germinate, irrespective of time.

- **germplasm**
entire array of cultivars in a crop species, related wild species in the genus, and hybrid between the wild and cultivated species.

- **gibberellin**
a plant hormone isolated from a fungus, used in promoting plant growth

- **gibbs free energy**
1. the thermodynamic function G=H-TS, where, H is enthalpy, T absolute temperature, and S entropy.
2. also known as free energy, gibbs function.

- **gibbsite**
is an aluminium rich mineral that accumulates in highly weathered soils. It has layers composed of $Al^{+++}$ ions in Octahedral coordination between two sheets of OH ions. Gibbsite crystals tend to be flaky because adjoining layers are weakly attracted to each other.

- **gilgai**
1. the microrelief of soils produced by expansion and contraction with change in moisture. Found in soils that contain large amounts of clay that swells and shrinks considerably with wetting and drying. Usually a succession of microbasins and microknolls in nearly level areas of microvalleys and microridges parallel to the direction of slope.
2. common feature of vertisols.

- **ginning**
a process of separating lint from cotton seed.

- **girth quotient**
the ratio of the girth of a stem at any given height to its

girth at breast height.

▪ **glacial drift**
a general term for the rock debris that has been transported by glaciers and is deposited either directly from the ice or from the meltwater, or melting, of the glacier. It may be heterogeneous or it may be assorted.

▪ **glacial soil**
soil formed by the grinding action of ice and snow brought down by glaciers in the upper regions of the Himalayas, found in Dehradun region and parts of Kashmir.

▪ **glaciofluvial deposits**
material moved by glaciers and subsequently sorted and deposited by streams flowing from the melting ice. These deposits are stratified and may be in the form of outwash plains, deltas, kames, and eskers.

▪ **glaebule**
a micropedological features consisting of a three-dimensional unit within S-matrix of the soil material, and usually approximately prolate to equant in shape, its morphology (especially size, shape, and/or internal fabric) is incompatible with its present occurrence being within a single void in the present soil material. It is recognised as a unit either because of a greater concentration of some constituent and or a difference in fabric compared with the enclosing soil material, or because it has a distinct boundary with the enclosing soil material.

▪ **glaisification**
clay enrichment of the accumulation horizons.

▪ **glass wool**
a material consisting of very fine glass threads, resembling cotton wool. Used for filtering and absorbing corrosive liquids.

▪ **glasshouse**
a building with glass walls and roof, for the cultivation and exhibition of plants under controlled conditions.

▪ **glei soil**
a soil horizon where poor drainage and intermittent water stagnation causes reduction of ferric compounds, so that the resulting ferrous compounds have a bluish-grey mottled appearance with stricky, clayey, structureless compact layer. The phenomenon is referred to as gleization.

■ **gley**
a portion of a soil altered by the water which persists within it, the iron has been reduced to the ferrous state, providing characteristic bluish or pale green colours throughout the soil mass or only in spots or streaks. It results in destruction of the structure, existence of asphyxiating conditions and impoverishment by leaching and reduced biological activity.

■ **gley horizon**
a horizon in a soil, characterised by the deposition of secondary hydrated ferric oxide.

■ **gley soil**
a soil developed under impeded drainage, and, in consequence, has a horizon of secondary hydrated ferric oxide. More generally, with a greenish grey shiny horizon, which may contain little iron.

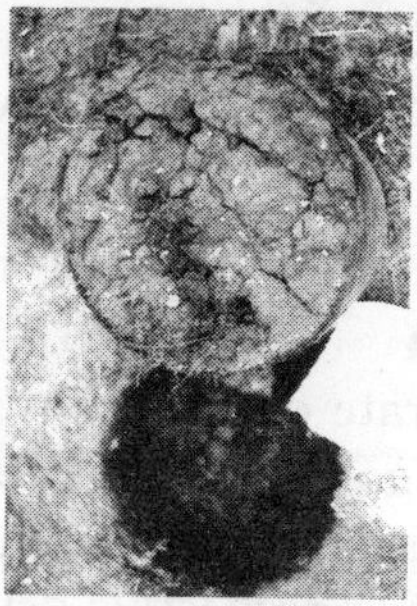

■ **global radiation**
the total of direct solar radiation and diffuse sky radiation received by a unit horizontal surface. Global radiation is measured by pyranometers.

■ **glutamic acid**
an amino acid, x-aminoglutaric acid, which is part of the vitamin folic acid and is important in nitrogen metabolism.

■ **glutathione**
a tripeptide, made of cysteine, glutamic acid, and glycine, found in various tissues and taking part in biological oxidations because it is capable of being alternatively reduced and oxidised, and also acts as a coenzyme to glyoxalase.

■ **gluten**
the protein-rich constituent of wheat grain, represented by albuminoids. Content lesser in soft wheats (8 per cent)' than hard wheats (26 per cent). Influences baking quality.

■ **glycophytes**
plants that are sensitive to a high salt concentration in the soil.

■ **gneis**
a coarse grained crystalline rock having foliated texture.

- **go to seed**
to mature beyond a usable condition.

- **go to stalk**
to grow vegetatively at the expense of seed or fruit.

- **golden syrup**
1. a sugar refinery term.
2. refined syrup partly inverted to avoid crystallisation.

- **gooch crucible**
a laboratory filter consisting of a shallow porcelain cup, the flat bottom of which is perforated with small holes over which a layer of asbestos fibres is placed.

- **good drainage**
the removal of excess surface or sub-surface water from land at such a rate that prevents damage to-growing crops or other desirable vegetation.

- **good will**
the benefits or value accrued to an existing business through its reputation and connections.

- **governed economy**
it is an alternative term for mixed economy.

- **gradient**
the steepness of a slope, expressed as a proportion between its vertical interval and its horizontal equivalent or as an angular measurement from the horizontal.

- **gradient profile**
a profile along a river, road or railway, showing mileage, slopes of different character and degrees of slope.

- **graduation**
the method or system of dividing a graduated scale, also, one of the equal divisions or one of the dividing lines in such a scale.

- **grafting clay**
also called Chaubattia clay-mud. A specially prepared clay used for plastering graft joints, found to be better and cheaper than grafting wax. Prepared by mixing 2 parts of soil having a clay fraction of 75 per cent with a pH of 6.0, 1 part of cowdung with an equal part of horse-dung soaked in water when cowdung has fermented, the mud is kneaded into a fine dough and mixed with the soil.

- **graham's law of diffusion**
the rate of diffusion of various gases are inversely proportional to the square roots of their respective densities.

Where, r, and d! represent the rate of diffusion and density of one gas, and *ti* and $d_2$ the corresponding values for a second gas.

■ **grain**

1. the fruit of large seeded grasses. Botanically a grain or kernel is a fruit containing one seed.

2. a unit of weight equivalent to 65 mg or 2/1000 of an ounce.

■ **grain cutans**

those cutans that completely coat individual sand and gravel grains.

■ **grain dryer**

mechanical device used for removing excess moisture from harvested grain, usually by passing unheated air through the grain.

■ **grain or particle size distribution curve**

curve obtained when the values of successive fractions of the soil are plotted against their corresponding settling velocities using logarithmic scale for the latter values. The most remarkable feature of the curve is the smooth character, the general form being sigmoid.

■ **gram negative**

bacteria that are not stained following the gram staining process.

■ **gram positive**

bacteria that are stained following the gram staining process.

■ **gramme**

one-thousandth part of a standard mass called the kilogram. The international kilogram is the mass of a cylinder of platinum iridium kept at the International Bureau of Weight and Measures at Sevres near Paris.

■ **grand period of growth**

in plants, the total period of enlargement of a cell, organ, or the whole plant, growth being slow at first, then increasing to a maximum, then slowing to zero at maturity.

■ **granites**

are igneous rocks. These are hard, massive, crystalline, coarse grained acidic rocks of plutonic origin. They are light coloured—mixture of pink and white or black and white —with felspar flakes and shining mica grains.

■ **granular**

consisting of grains or granules, appearing as if made of granules.

**granular fertiliser**
a fertiliser composed of particles of roughly the same composition, about 1/10 inch in diameter.

**granular pesticide**
a pesticide chemical mixed with or coating small pellets or a sand-like material.

**heteroauxin**
a substance, chemically identical to B-indo-leacetic acid, extracted from fungi and urine, which acts as a growth promoting hormone.

**heterocalcifere**
soil in which the calcium carbonate does not come from the soil itself but from elsewhere.

**heterokaryosis**
the association of genetically diverse nuclei in an organism capable of propagation. This phenomenon is common in fungi.

**heterometry**
the character of a material which is composed of grains of different sizes.

**heteropoly acid**
complex acids of metals, whose specific gravity is greater than 4, with phosphoric acid, example: phosphomolybdic acid.

**heterotroph**
a microorganism which is unable to use $CO_2$ as its sole source of carbon and which requires a range of exogenous organic compounds for growth and reproduction. A heterotroph which obtains energy from light is referred to as a photoheterotroph. A microbe which lives chemotrophically is referred to as a chemoheterotroph.

**heterotrophic**
a group of organisms which obtain carbon for synthesis from other organic matter or proteins. All animals and fungi, most bacteria and a few flowering plants are strongly heterotrophic, requiring organic food substance from which they make most of their own organic constituents.

**heterotrophic bacteria**
are defined as those which are not self sustaining. They are unable to form protein and carbohydrates from inorganic carbon and nitrogen.

**heterozygote**
an individual produced by union of two germ cells that contain unlike genes for a given character, either both

genes of an allelomorphic pair or two different genes of an allelomorphic series.

- **hibernation**
reduction of biologic activity by organisms during winter, or, more generally, during cold periods.

- **hidden erosion**
slow mechanical erosion of loose soil.

- **hidden hunger**
1. describes a plant that shows no obvious deficiency symptoms, yet the nutrient content is not sufficient to give the top profitable yield.
2. Fertilisation with the sure rate rather than the bare economic optimum for an average helps to obtain the top profitable yield.

- **hidden solonchak**
contains salts precipitated only within the soil, occurs in the case of deep water table.

- **high**
in meteorology, any area in which atmospheric pressure is higher than in the surrounding areas. The term is used in .a more general sense than an anticyclone, which is centred on a closed isobar.

- **high acre value crops**
are crops such as fruits, vegetables and tobacco with which the acre value of the produce is quite high.

- **high calcic products**
materials of which 90 per cent or more of the total calcium and magnesium oxide content consists of calcium oxide.

- **high duty of water**
an irrigation water requirement where a given quantity of water will serve a relatively large area of land.

- **high energy bond**
a bond that yields a large (at least 5 Kcal mole) decrease in free energy upon hydrolysis.

- **high order interaction**
the interaction between three or more variables.

- **high volume spraying**
spraying when the spray material deposited over surface of leaves is in large drops, when they spread over surface as film, accumulate at edge, finally runoff on to ground, normally obtained by spraying over 50 gallons of spray

liquid per acre.

- **highly volatile**
a liquid that quickly forms a gas or vapour (evaporates) at room temperature.

- **hill**
uplands of less elevation than mountains in the same area.

- **hill dropping**
a method of sowing in which seeds are dropped at fixed spacing and not in a continuous stream. Thus the spacing between plant to plant in a row is constant.

- **hill placement**
application of fertiliser either in bands or in local areas near the plants like cotton, cabbage, tomato, etc.

- **hill reaction**
the reaction by which isolated illuminated chloroplasts bring about the reduction of certain substances with accompanying evolution of oxygen. The reaction involves part of the normal light reaction of photosynthesis. Electrons from water involved in non-cyclic photophosphorylation are used to reduce the added substance. It provided support for the idea that a light reaction preceded reduction of carbon dioxide in photosynthesis.

- **hilum**
scar on seed coat, marking point of former attachment of seed to funcile.

- **histicepipedon**
a surface horizon that contains more than 20 to 30 per cent organic matter, depending on clay content, and is water saturated for 30 days at some time of the year unless artificially drained. It is thinner than 30 cm if drained or 45 cm if not drained.

- **histidine**
an aminoacid with an indole nucleus $C_6H_9N_3O2$, one of the components of humus which is not precipitated by the acids, derived from proteins, can be poisonous to the soil.

- **histogram**
1. a graph of computations furnishing the data for plotting a histograph a bar diagram representing a frequency distribution.
2. a graphical representation of a frequency distribution by means of rectangles whose widths are the class intervals and whose heights are the corresponding: frequencies.

- **histograph**
a map or chart of a river drainage or sewer system,

upon which a series of time lines are placed. These time lines give the time of transit of water originating on a time line, to flow down to the outlet of the system.

- **histology**
  the science which treats of the detailed structure of animal or plant tissues, microscopic morphology.

- **historical school**
  a school of economic thought which wished to supersede the deductive method of the classical school by observation from past history. It favoured art inductive study of the economic system. Its chief architects were primarily Germans of the mid-nineteenth century.

- **hoe**
  A hand tool consisting of a handle of variable length depending upon the region and a light cutting blade at right angles to it. Used for light work such as weeding and mulching the top soil in between crop rows in garden or field.

- **hluviation cutans**
  cutans formed by movement of the cutanic material in solution or suspension and subsequent deposition. A sharp boundary between the cutanic and non-cutanic material is indicative of an illuviation cutan. They are always associated with conducting voids unless they have been disturbed and disrupted by subsequent mass movement or differential movement of the soil material.

- **hoe out**
  to remove weeds from a field with a hoe.

- **hofmeister series**
  an arrangement of anions or cations in order of decreasing ability to produce coagulation when their salts are added to lyophilic sols.

- **holard**
  the whole of the water in the soil.

- **hold back carrier**
  synonym: Hold-back agent. Inactive isotopes of a particular nuclide employed to retard the co-precipitation or absorption of a radionuclide.

- **holding**
  agricultural land that is being cultivated by an owner, tenant, or manager as one unit. Sometimes a number of farms are worked as one holding.

- **holding company**
  a company that owns the

majority of stock or securities of one or more other companies for purpose of control rather than of investment. A company which exists only for this purpose is a pure holding company.

- **holo calcaire**
  soil in which the calcium carbonate comes from the soil itself.

- **holocoenotic principle**
  principle that environment acts as a whole because of lack of barriers to the interaction of its component factor.

- **holoenzyme**
  a complete enzyme, made up of evapo-enzyme and a prosthetic group.

- **holophytic**
  1. capable of synthesising complex molecules from simple ones, thus storing chemical energy, and building new tissues from them.
  2. This refers especially to photosynthesis.

- **holophytic nutrition**
  nutrition involving photosynthesis from simple inorganic chemical substances, as in green plants.

- **homeostasis**
  the buffering capacity of an ecosystem that allows it to resist perturbations, the tendency' of a biologic system to remain at or return to normal, after or during an outside stress, resistance to change and a tendency toward equilibrium, maintenance of constant level conditions in the face of a varying external environment.

- **homobium**
  a self supporting association of a fungus and an alga, as lichens.

- **homogamy**
  condition in which male and female parts of flower mature simultaneously.

- **homogenous**
  pertaining to a substance having uniform composition or structure.

- **homoionic clay mineral**
  a mineral in which all the negatively charged sites are neutralised by a single type of exchangeable cation.

- **homoiothermal**
  having constant interval temperature, often maintained above that of the environment.

- **homophyllous**
  having foliage-leaves all of the same kind.

- **homoplastic**
  of the same structure and

manner of development, but not descending from a common source. Describing similarity due to convergence or parallel evolution.

- **homozygote**
  an individual produced by union of two germ cells that contain like genes for a given character.

- **hoof and horn meal**
  processed, dried, and ground hoofs and horns.

- **horizonation**
  includes the proanisotropic process and conditions by which initial materials are differentiated into soil profiles with many horizons.

- **horizontal amalgamation**
  a merger of firms at the same stage of production, as for example, a group of firms engaged in cotton spinning.

- **horizontal bench terrace**
  a type of bench terrace which has no measurable slope from back to front of the bench.

- **horizontal eluviation**
  a horizon.

- **horizontal illuviation**
  b horizon.

- **horizontal silo**
  a silo built with its long sides parallel to the ground surface, with openings on either side to facilitate filling.

- **horizontal zonality**
  the zonation of soils over broad areas in the climatic zones from the poles to the equator is called horizontal zonality.

- **hormesis**
  the theory that small doses of radiation can induce beneficial biological processes and are healthful.

- **hormonal weedicides**
  1. hormones, which when applied to growing weeds, bend, control, or twist the stem and leaves, sometimes make the stem thicken and split and in most cases arrest its growth. Application results in the death of plants.
  2. Also selective killers, killing some plants outright even with a trace of the hormonal substance, and not injuring others, useful in the eradication of weeds, applied before or after emergence of crop.

- **hormone**
  a chemical messenger within the body that is secreted by one type of cell and acts on another type of cell.

- **hormones plant**
  1. are regulators produced by plants, which in low concen-

tration regulate plant physiological process. Hormones usually move within the plant from a site of production to a site of action.

2. Also called phytohormones.

- **horological**
said of a flower which opens and shuts at a definite time of day.

- **horse manure**
dried horse excrement used as manure.

- **horse power**
a British unit of power representing a rate of work of 550 ft. 1 b. wt. per second. One horse power is equal to 746 watts.

- **host specificity**
the intimate relationship between a parasite and its host which limits the host range of the parasite to the one particular (or closely related) host species.

- **hot laboratory**
laboratory equipped for the manipulation of chemical processing of highly radioactive materials.

- **hot mix liquid fertiliser**
a liquid mix fertiliser manufactured by the neutralisation of phosphoric acid with anhydrous or aqueous ammonia. Urea or a urea-ammonium nitrate solution is generally required as supplemental nitrogen material, potassium chloride is usual source of K.

- **hot money**
it refers to the short-term capital movement from one country to another seeking either a higher rate of interest or safety of capital irrespective of the rate of interest.

- **hot water treatment**
done for control of internal seed borne fungus. Soak the seed in water for 4 hours and dipped in hot water for one minute followed by dipping in warm (54°C) water for 10 minutes.

- **hue**
the wavelength of light of a pure colour that gives a colour its name—such as red or blue—independent of its saturation or brightness.

- **hulling**
process employed for converting paddy into rice, carried out either as simple indigenous method using manual labour only, or by the use of power driven machinery, the former accounting for three-fourths of the rice produced in India.

**humid**

1. an adjective applied to regions or climates where precipitation is adequate in amount and occurs at such times that agriculture can be carried on without irrigation, but where drainage may be required to remove excess moisture. This denotes a climate with sufficient precipitation to support a forest vegetation.
2. containing, or characterised by, sensible moisture, moist or damp atmosphere.

**humid climates**

1. are those where precipitation equals or exceeds the evaporative demand of the atmosphere for most periods of the year.
2. The large fraction of cloud cover which reduces incident solar radiation cuts down the evaporation rates, and helps to maintain a favourable water balance for the vegetation, even if the total rainfall is low.

**humidity**

1. to add water vapour to the atmosphere.
2. to add water vapour or moisture to any material.

**humidity**

the term refers to the amount of water vapour content of the atmosphere.

**humidity index (IH)**

the ratio of water surplus (S) to potential evaporation (PE) or water need (S/PE) usually expressed as a percentage.

**humidity mixing ratio**

defined as the amount of water vapour in grams mixed with one kilogram of dry air. Common values are in the order of 4.30 gm per kg. For each temperature there is a corresponding saturated humidity mixing ratio.

**humification**

the process-of decomposition of organic matter in soil. During humification, the major part of the inorganic elements in the original organic matter is mineralised into simple inorganic compounds and passed into the soil solution.

**humified soil**

soil in which the organic matter has been greatly transformed, and the humus of which thus has a high nitrogen content. Humified soils have a C:N ratio between 8 to 12 but poorly humified soils have C:N ratio in the range of 20-30.

**humilluvic material**

diagnostic organic material. Illuvial humus that accumulate after prolonged cultivation of some acid organic soils.

- **humm**
one of the constituents of humus which is insoluble in alcohol, water and sodium carbonate. It is not destroyed by acetyl bromide.

- **humo ferric complex**
apseudosoluble electronegatively colloidal association of ferric hydroxide (electropositive) and humic acid (electronegative). The resulting hydrosols do not flucculate in the presence of the electrolytes of soil solutions, and constitute a form in which iron migrates.

- **humo phosphate**
a complex of alkaline humate and mono-calcium phosphate. The cation $Ca^{+}+$ forms the link between humic acid and $PO_4^{3'}$. Humophosphate is weakly soluble in water.

- **humus**
decomposed plant and animal material that is part of the soil. Humus is a product of composting.

- **humus calcareous soil**
soils rich in humus and calcium carbonate.

- **humus formation**
the transformation of plant and animal debris by the micro-organisms *in situ* results in humus formation.

- **humus layer**
the top portion of the soil, which owes its characteristic features to its content of humus. The humus may be incorporated or unincorporated in the mineral soil.

- **humus nucleus**
a large hypothetical molecule resulting from the linking of lignin and proteins, which forms the basis of humus.

- **humus plant**
a flowering plant, often poorly provided with chlorophyll, and which grows into deep humus. Its roots then form a mycorrhiza with a fungus.

- **humus theory**
first stated by Aristotle, that plants absorb food from the soil in an elaborated form.

- **husk**
the outer coating of various seeds.

- **hybrid**
hybrid plants are the result of a cross-pollination, that is, the breeding of two plants that differ genetically in some way, the two plants can be of different varieties or even species.

- **hybrid vigour**
  the increased vigour of growth in a cross between two genetically different parents as compared with growth in either of the parents.

- **hybridisation**
  a method of crop improvement in which two or more plants of unlike genetical constitutions differing in one or more characters are crossed together.

- **hydathode**
  water-extracting gland occurring on the edges or tips of leaves of many plants.

- **hydrase**
  an enzyme which can add or remove water, with hydrolysis.

- **hydrated electron**
  an electron released during ionisation of a water molecule by water and surrounded by water molecules oriented so that the electron cannot escape. Also known as Aqueous Electron.

- **hydrated or air slaked lime**
  a dry product consisting chiefly of the hydroxide of calcium and the oxide and hydroxide of magnesium.

- **hydration**
  the chemical union of a compound with water.

- **hydraulic**
  1. of, involving, moved by, or operated by a fluid, especially water, under pressure.
  2. able to set and harden under water, as Portland cement.
  3. of or relating to hydraulics.

- **hydraulic agitator**
  a device which keeps the tank mix from setting out by means of water flow under pressure.

- **hydraulic conductivity**
  the coefficient 'K' in Darcy's law $v = Ki$ where v is the velocity of seepage and i is the hydraulic gradient. Value of K depends on properties of the fluid, such as viscocity as well as that of the porous medium, and reflect any interactions of the fluid with the porous medium such as swelling of soil and its attendant reduced porosity.

- **hydraulic equilibrium of water in soil**
  the condition for zero flow rate of liquid or film water in the soil. This condition is satisfied when the pressure gradient force is just equal and opposite to the gravity force.

- **hydraulic friction**
the resistance to flow exerted on the perimeter or contact surface between a stream and its containing conduit, due to the roughness characteristic of the confined surface, which induces a loss of energy. Energy losses arising from excessive turbulence, impact at obstructions, curves ad-dies, and pronounced channel changes are not ordinarily ascribed to hydraulic friction.

- **hydraulic gradient**
the slope of the hydraulic grade line, the rate of change of pressure head, the ratio of the loss in the sum of the pressure head and position head to the flow distance. For open channels, it is the slope of the water surface and is frequently considered parallel to the invert. For closed conduits under pressure, it is the slope of the line joining the elevations to which water would rise in pipes freely vented and under atmospheric pressure. A positive slope is usually one which drops in the direction of flow. In dikes, embarkment earth, dams and earthen structures retaining water, the line of saturation in the earth.

- **hydraulic head**
the distance between the respective elevations of the upstream water surface (headwater) above and the downstream surface water (tailwater) below a hydropower plant. For non-turbulent flow of water in soil the velocity head is negligible. In unsaturated soil a porous cup must be used for establishing hydraulic contact between the soil water in a manometer. Hydraulic head has the dimension of length (L).

- **hydraulic permeability**
the rate of discharge through a unit cross-sectional area of soil normal to the direction of flow when the hydraulic gradient is unity.

- **hydraulic radius (r)**
is the ratio between the cross-sectional area of the stream and its wetted perimeter. It is an important parameter in determining the velocity of water flowing in channels. Velocity of flow in channels is directly proportional to the square root of the hydraulic radius. The hydraulic radius decreases with the increase in wetted perimeter, for a given cross-sectional area.

- **hydraulic sprayer**
a machine which applies pesticides and other chemicals at

high pressure and volume to deliver the contents to the target. Same as high-pressure sprayer.

- **hydraulic structure**
any engineering structure for the purpose of water control.

- **hydric**
characterised by an abundant supply of moisture.

- **hydro**
a prefix indicating a connection with water.

- **hydrocapric**
said of aquatic plants which ripen their fruits under water, after pollination has taken place in the air above.

- **hydrocarbon**
an organic chemical compound consisting only of carbon and hydrogen atoms in the gaseous, liquid or solid phase.

- **hydrocyanic acid (HCN)**
a poison, also called prussic acid, produced as a glycoside by several plant species: especially sorghums.

- **hydrogen bacteria**
a number of bacteria including *Bacillus Jiydrogenes, Pantotrophus* spp. and *Hydrogenomonas* spp. which obtain their energy by oxidising hydrogen to water.

- **hydrogen bond**
a intermolecular interaction formed between a hydrogen atom covalently bonded to an electronegative atom (such as oxygen or nitrogen) and a second electronegative atom that serves as the hydrogen bond acceptor.

- **hydrogen sulphide**
a colourless poisonous gas with a smell of bad eggs. It is formed by the decomposition of organic matter containing sulphur and occurs naturally in some mineral waters. It is prepared by the action of dilute acids on sulphides of metals, used in chemical analysis.

- **hydrogenation of oils**
artificial hardening of liquid animal and vegetable oils by the action of hydrogen. Liquid fats and oils contain a high percentage of liquid triolein, $C_{51}Hi_{M}O_{6}$, which may be converted into a solid tristearin, $C_{57}HnoO_{6}$, by the action of hydrogen in the presence of a finely divided nickel, catalyst, the result being a hard fat of higher melting point.

- **hydrogenic soil**
soil developed under the influence of water standing within the profile for considerable periods, formed mainly in cold, humid regions.

▪ **hypothesis of felback**

a hypothesis about the nature of humic substances. Humic substances consist of a microbially produced central unit of 4-pyrone units linked together in a chain by methylene bridge at the 2.6-positions. Phenolic and amino-acid units are thought to be attached to the central structure.

▪ **hypothesis of flaig**

a hypothesis about the nature of humic substance which are thought to be the end-product in the sequence starting with the degradation and demethylation of lignin to substituted polyphenols, followed by oxidation to quinones, and finally the condensation of these quinones with amino-acid units into humic acid or its immediate precursors. It requires the plant lignin as a source of the elements that form the main structural unit of humic acid.

▪ **hypotonic**

designating a solution with an osmotic pressure less than that of a specified other solution, the latter being hypertonic. When separated by a semipermeable membrane, water is lost by osmosis from the hypotonic to the hypertonic solution.

▪ **hypsometer**

an instrument for measuring the height above sea level by determining the atmospheric pressure through observation of the boiling point of water, since this varies with pressure of the atmosphere.

▪ **hysteresis**

1. a soil drying out has a higher equivalent tension at the same moisture content than when it is being wetted. This is referred to as hysterisis effect which is particularly pronounced at the higher moisture contents.

2. the phenomenon of dependence of the state of a system on its previous history.

3. a lag in an effect with respect to its cause. It results in the phenomenon developing differently in one direction than in the other.

■ **hythergraph**
a climatic diagram plotting temperature against humidity or precipitation. It summarises broad climatic differences in relation to human activity.

■ **ice fog**
a suspension of numerous minute ice crystals in the air, reducing the visibility at the earth's surface.

■ **ideal soil**
a soil which conforms to the idea one has of the normal evolution through the action of the factors present under natural conditions without disturbances.

■ **ideotype**
a plant with its morphology ideally suited for getting maximum produce under a particular condition or set of conditions.

■ **idle land**
land which has been cultivated, but put to disuse, abandoned land, fallow land.

■ **ignition loss**
the loss of weight from an oven dried soil sample, on burning.

■ **illite**
a 2:1 layer silicate that is either primary mineral inherited from the parent rock or intermediate products of weathering. It is often mixed with montmorillonite. Illite layers have more negative charge than montmorillonite layers because one-fourth of the tetrahedral positions in illite are occupied by $Al^{+++}$ instead of Si+$^{++}$+. The negative charge is offset by K+ ions held between the layers. The K+ ions fit into the holes in the tetrahedral sheets and hold adjacent layers together too tightly for water to enter. The illite lattice is therefore nonexpanding. The $K^{+}$ ions in the lattice are not available to plants but gradually become available as they are released by weathering.

■ **illuvial horizon**
horizon that has received material in solution or suspension from some other part of the soil.

■ **illuviation**
the process of deposition of soil material removed from one horizon to another horizon of the soil, usually from an upper horizon to a lower horizon in the profile.

■ **imbibition**
the process of absorption by which solids, especially organic substances in the gel

state, take up fluids, and in the process, swell. For example, absorption of water by seeds, wood, gelatine. One of the means of absorbing water by roots and uptake of water by seeds and the splitting of seed coats.

- **imbibitional moisture**
moisture over and above needed to satisfy the hygroscopicity of soil particles. Turns soil visibly moist by the colloidal clay imbibing moisture, clay particles increase in size and swell. Held fast and tenaciously by clay, but available to plants for absorption.

- **immature**
trees or stands that have grown past the regeneration stage, but are not yet mature.

- **immature staple**
refers to cotton which has been picked and baled before the fibre has been completely developed, weakened staple of inferior value.

- **imminent hazard**
a situation which exists when the continued use of a pesticide would likely result in unreasonable, harmful effects on the environment.

- **immobilisation**
the conversion of an element from inorganic to organic combination in microbial or plant tissues. This has the effect of rendering unavailable (and usually not readily soluble) an element that previously was leachable or directly available to plants.

- **immune**
1. protected against a disease.
2. immunise the process of giving protection from disease through the body's own immune system.

- **impact effect**
it refers to the immediate effect of a change in demand when supply cannot be immediately readjusted to the new situation.

- **impeded drainage**
condition in which downward movement of gravitational water is hindered.

- **imperfect humus**
humus in the initial stages of formation. It is not well mixed with soil but resting on it.

- **imperfect market**
it is a market characterised with the existence of imperfect competition. In such a market, the following conditions of perfect competition are not

fulfilled:
a. the commodity is homogenous.
b. there is a large number of both buyers and sellers.
c. buyers and sellers are in close touch with each other.
d. the commodity must be transferable.
e. there must be no favourable treatment of some buyers or discrimination against others.
f. only a few markets are perfect in actual practice. Most of the markets are imperfect.

- **imperfect oligopoly**
this refers to that situation where the production of a commodity is in the hands of only a few producers and their products are differentiated from each other. The extent of differentiation is generally small and is mainly due to the advertisement and publicity conducted by the rival producers. Each producer attempts to make the consumers believe that his product is superior to that of others. Imperfect oligopoly often leads to price war or advertisement wars.

- **imperfect or limited monopoly**
a single-firm industry where the cross-elasticity of demand between the product of the firm and that of other commodities in the market is small though it is above zero.

- **impermeability**
the property of rock, earth or other material which prevents perceptible movement of water. Even though the material allows capillary movement, it still may be impermeable.

- **impervious**
1. a term applied to a material through which water cannot pass or through which water passes with great difficulty.
2. resistant to penetration by fluid or roots.

- **impervious bed**
a bed or stratum through which water will not move under ordinary hydrostatic pressure.

- **imperviousness coefficient**
the ratio expressed decimally, of effectively impervious surface to the catchment area under consideration.

- **implement**
a tool or device that attaches to or is towed by a tractor.

- **implements for carrying soil**
they are wheel barrows, keni or levelling karah, leveller and

clod crusher and bund formers.

- **impoverished soil**
soil which has been continuously cropped and has had little or no care so that it is incapable of supporting satisfactorily an economic crop.

- **improved land**
land that has had its value increased either because it has been reclaimed from waste for agriculture or because it has had its yield potential increased by such methods as drainage or levelling.

- **improved seed**
seed which is improved over local variety.

- **in line**
description of the processing of a liquid within a pipe system that contains it.

- **in season**
that part of the year when a product is normally harvested and is thus cheaper and more plentiful.

- **in situ water harvesting**
a technique of water harvesting where water is stored in the soil profile itself.

- **in vitro**
literally in glass, describing biological experiments in laboratory apparatus rather than in the living organism. Cell tissue cultures are example of *in vitro* experiments.

- **in vitro protein synthesis**
the incorporation of amino acids into polypeptide chains in a cell-free system.

- **in vivo**
literally 'in life', describing processes that occur within the living organism. A group of organisms so highly inbred as to be genetically identical, except for sexual differences. Brother sister matings have been maintained for 20 or more generations.

- **inbreeding**
mating of close relatives, which reduces genetic diversity, often leading to expression of deleterious recessive characteristics and reduction of fitness in the offspring.

- **inbreeding coefficient**
a measure of the intensity of inbreeding.

- **inch**
a unit of length equal to one twelfth of a foot. This relationship was established by the Congress on July 28, 1896.

- **incineration**
burning the sludge to remove water and reduce remaining residues to ash.

- **incipient erosion**
the early stages of erosion, especially with reference to gully erosion.

- **incipient plasmolysis**
the condition when about half the cells of a tissue are plasmolysed. It is in this condition that the osmotic pressure of cell-sap is usually measured.

- **incipient wilting**
small loss in turgidity during warm weather, even when soil is moist. May not be visible to the naked eye. Usually reversible at night.

- **included water**
water which is in the clods of the soil.

- **income effect**
the effect on the purchases of the consumer caused by changes in income with prices of goods remaining constant.

- **incomplete soil**
soil which has not arrived at maturity, which does not have the characteristic horizons that it presumably ought to have.

- **incorporate into soil**
the mixing of a pesticide into the soil by mechanical means.

- **increasing marginal cost**
a situation in which the extra cost of producing each additional unit of output rises as production is increased in the short run.

- **increasing returns to scale**
a situation in which output increases more than proportionately as all factor inputs are increased in the long run.

- **increment**
when the output or the input of a specified quantity increases.

- **incrustation**
a coating of iron carbonate, or iron compounds on the surface of some algae.

- **incubation period**
the time interval between the exposure to a disease and the appearance of the first symptoms or between the time of inoculation and visible growth

of bacteria in the media.

- **index**

1. an indicator, usually numerically expressed, of the relation of one phenomenon to another.

2. an indicating part of an instrument.

- **indicator host**

a plant that shows symptoms of viruses that are latent in other hosts.

- **indicator line**

a fixed line in an experimental area along with periodic observations are made.

- **indicator plant**

a native plant that indicates, in general, and often in specific manner, the nature of soil conditions with regard to moisture and salinity. Dominant species are the most important indicators of such conditions.

- **indicator plot**

a small plot kept under detailed observation within a larger experimental area, *e.g.* when testing regeneration methods.

- **indivisibility**

it means the indivisibility of certain factors of production into smaller units. For example, a big plant cannot be divided into half and produce half its maximum output at half its previous costs. The plant can, however, work at 50 per cent of its capacity, but then its operating cost will have to be spread over 50 per cent of its maximum output, raising thereby the unit cost of production. Indivisibility is an important factor in producing economies of large-scale production. Indivisible factors can be most economically employed when output is large, they work less economically when output is low, because they cannot be divided into smaller units.

- **indo-gangetic alluvium**

this is by far the most important soil group of India contributing the largest share to the agricultural wealth of the country. The main features of these soils are derived from the deposition of silt by the numerous tributaries of the Indus, Ganges and Brahmaputra rivers. The soils vary in consistency from drift sand to loams and from fine silts to stiff clays. The soils as a whole are young and clay movement to the substrata is minimal. Their fertility is variable and in some areas they are poorly leading to the accumulation of salts.

■ **induced crusts**
are compacted soil at the soil surface about 1 cm thick caused by raindrop splash.

■ **induced pans**
those soils where the restrictive layer is the result of a recently applied compacting force, such as implement traffic or trampling, upon a soil that had, under virgin conditions, physical properties favourable to the penetration of roots and water.

■ **induced radioactivity**
radioactivity produced in certain materials as a result of nuclear reactions, particularly the capture of neutrons, which are accompanied by the formation of unstable (radioactive) nuclei. In a nuclear explosion, neutrons can induce radioactivity in the weapon materials, as well as in the surroundings.

■ **inducible enzymes**
enzymes whose rate of production can be increased by the presence of inducers in the cell.

■ **industrial sewage**
sewage in which industrial wastes predominate.

■ **industrial wastes**
the liquid wastes resulting from the processes employed in industrial establishments.

■ **ineffective rainfall**
is that portion of the total annual or seasonal rainfall which is lost by surface run-off, deep percolation beyond root zone, the moisture remaining in the soil after the harvest of the crops and is not useful for the next crop. The period considered here is from the start of the- first tillage operation till the harvest and not only the actual crop growing period.

■ **inert ingredient**
any ingredient of a pesticide product that is not the active ingredient. Inert ingredients are not displayed on the product label and are often highly toxic. They are added to increase the potency of the active ingredient, or to act as a carrier.

■ **infection**
the invasion of and growth within an organism by pathogenic microorganisms with the consequent development of disease.

■ **infective**
caused by infection or capable of causing infection.

■ **infective dose (ID)**
the number of microorgan-

isms required to infect a host.

**infective stage**
stage in the development of an egg or larva at which it is capable of becoming established in the definitive host.

**infest**
to overrun the surface of a plant, or to *be* dispersed through soil or other substrate.

**infestation**
the lodging, development, and reproduction of vermin within a defined area.

**infestation spot**
an infestation of uniform degree, limited to a definite small area, such as part of a stand.

**infield**
the infield is the portion of the field in fair territory which includes area normally covered by infielders. The infield is normally fenced, manured, and cultivated.

**infiltration capacity**
1. it describes the disappearance of water at the surface of the soil into the profile. When there is no time lag between the arrival of a water drop and its penetration through the soil surface, the infiltration rate is equal to the water application rate.
2. the constant rate at which water can percolate into a soil.

**infiltration coefficient**
the ratio of infiltration to precipitation.

**infiltration complex**
a group of areas which have comparable similar physical conditions of soil, slope, litter, and cover, and which have a similar infiltration rate.

**infiltration index**
rate of infiltration calculated from records of rainfall and runoff. There are several different indices, each calculated by a different method. A common index is the average rainfall intensity for a given storm above which the mass of rainfall equals the mass of runoff.

**infiltration rate**
1. the maximum rate at which a soil, in a given condition at a given time, can absorb water.
2. also, the maximum rate at which a soil will absorb water impounded on the surface at a shallow depth when adequate precautions are taken regarding border or fringe effects.
3. defined as the volume of water passing into the soil per unit of area per unit of time.

- **infiltration velocity**
the volume of water moving downward into the soil surface per unit of area per unit of time. The maximum infiltration velocity is the infiltration rate.

- **infiltration volume**
the quantity of rainfall, expressed in inches depth over a given area, that passes below the root zone of plants and ultimately reaches the zone of saturation.

- **infiltration/inflow**
the uncontrolled inward leakage of air through cracks and gaps in the building envelope, especially around windows and doors.

- **infiltrometer**
a device by which the rate and amount of water percolating the soil is determined by measuring the difference between the amount of water applied and that which runs off. The water may completely cover the ground inside an infiltrometer ring or may be applied by a sprinkling mechanism (rain simulator) which provides a rather uniform sprinkling of water to a prescribed area at prescribed rates and size drops (impact) auxiliary apparatus consisting of a rain gage (either total or intensity) and a catchment basin or receptacles in which surface runoff is measured either as to rate or total flow. Infiltration of amount absorbed by soil is usually expressed in inches (of water) per standard interval of time.

- **inflation demand pull**
it refers to that rise in the price-level which takes place because consumers and investors with rising income compete with for a relatively limited supply of available goods.

- **inflationary gap new**
it refers to the excess in average hourly earnings of workers over the increase-in output per man. Either this gap can be closed by a reduction in the rate of wage increase or an increase in output. Per

man, an hour.

- **influence area**
  the surface area surrounding a well, or group of wells, during pumping, within which the water table or other piezometric surface is lowered by withdrawal of the water. The area varies in extent with the rate and duration of pumping.

- **influent stream**
  a stream type, common in an arid climate, that loses water into the ground, thus recharging the ground-water, unlike the effluent streams that are usual in wetter countries.

- **influx**
  the process of flowing in of ions by a plant root.

- **inherent resistance**
  quality of a plant to resist attack of pest or disease, probably associated with some particular characters of structure or composition.

- **inherit profile characteristics**
  are those inherited from the parent material and hence geological in nature *e.g.* the presence of quartz mica and feldspar in soils.

- **inheritance**
  the reception or acquisition of characters or qualities by transmission from parent to offspring.

- **inherited character**
  the character derived by individuals from one or the other of the parents, the expression of which is determined by a particular gene or genes.

- **inner capillary water**
  that fraction of capillary water which is held at tensions between 15 and 31 atmosphere. It is difficultly available to plant roots. The plant begin·to wilt.

- **inner space and apparent osmotic volume**
  1. the space in a cell or tissue to which ions cannot penetrate by free diffusion.
  2. That volume of a plant cell or tissue which will not allow free diffusion of ions to take place.

- **inning**
  the act of enclosing or reclaiming land, the land enclosed, especially marshland.

- **innovation**
  the successful application of new inventions. It is a source of economic growth.

- **inoculum**
  a substance or organism that is introduced into surroundings suited to cell growth.

▪ **inorganic**
describes material that is of mineral origin. Specifically, describes chemical compounds that do not contain carbon and hydrogen.

▪ **inorganic nitrogen**
nitrogen in combination with minerals, not in animal or vegetable form.

▪ **inorganic soil**
a mineral soil, a soil in which the solid material is mostly rock minerals in contrast to organic soil such as peat.

▪ **input (gross expenditure)**
includes
i). value of hired labour paid in cash or kind or both.
ii). wages of the farmer and his family members.
iii). value of seed, manure and other cost expenses (farm produced seed and manure are also included).
iv) rent paid for land whether rented or owned.
v). depreciation of livestock and deadstock.
vi) interest on working capital.
vii) the energy given to the machine or the work done on the machine.

▪ **input output access**
a measure of relative advantage of a region as compared to other regions in their access to markets and/or inputs. Greater input-output access indicates a higher growth potential of the region.

▪ **input-output analysis**
it is also known as inter industry analysis. It refers to that analysis which deals in great detail with the interrelationships between an industry's or an economy's given output of goods and services and the volume of goods and services needed to achieve that given output of goods and services.

▪ **intraspecific**
the interactions that occur within the individuals of the same species.

▪ **intrinsic viscosity**
the ratio of a solution's specific viscosity to the concentration of the solute, extrapolated to zero concentration.

▪ **invasion**
the migration of organisms (usually plants) from one area to another and their establishment in the latter.

▪ **inventory**
an inventory is a list of assets and liabilities which are claims or debts against the business in other words it is a detailed

list of farm properties, with values assigned.

- **inverse yield nitrogen law**

Wilcos put forth this law which expresses the fact that the power of growth or yielding ability of any crop plant is inversely proportional to the mean nitrogen content in the dry matter. The relative yielding ability of the crop plants depends upon the relative abundance of nitrogen in their dry matter but the relationship is inverse, a crop plant with high mean per cent nitrogen content in dry matter has less dry matter production potential than a crop plant with low per cent nitrogen content. With the aid of this law, it is possible to determine the relative yielding capability of crop plants and their varieties. This law is expressed by the equation:

$Q=318/N$

This law tends strong support to the theory of evolution. It also fixes the penultimate yield expected from any crop plant under an ideal environment.

- **invert emulsion**

are referred to the element that the water is dispersed in oil rather than oil in water. Oil forms the continuous phase with the water dispersed therein, usually a thick, mayonnaise like mixture results.

- **involuntary unemployment**

this occurs when an individual is out of work though no fault or wish of his own. As against this, 'voluntary unemployment' refers to an individual who prefers not to work even though suitable work is available to him.

- **iodine**

a sanitising agent similar in use to chlorine. Not widely used because relatively weak, cannot be stabilised, and tends to turn water greenish-yellow or even brown. The plant uptake closely follows the soil iodine content. The iodine content of plants usually varies within the ranges 0.1 to 1.0 ppm in dry matter. The critical deficiency limit reported in literature seems to lie around 0.05 to 0.07 ppm in dry matter. Visible evidence of iodine deficiency occurs most frequently in newborn farm animals in goitrous areas. The iodine content of common fertilisers varies usually from traces to about 20 ppm and up to several thousands ppm in organic manures containing seaweeds.

- **ion exchange**
a reversible process in which ions are released from an insoluble permanent material in exchange for other ions in a surrounding solution, the direction of the exchange depends upon the affinities of the ion exchanger for the ion present, and the concentrations of the ions in the solution.

- **ion exchange resins**
synthetic resin beads made for ion exchange. Ions which are held by electrostatic forces to electrically charged groups on the surface of the bead interchange with similarly charged ions in solution in water being treated. For obtaining high quality distilled water, cation exchange resins are used first and then anion-exchange resins, but mixed-bed ion exchange is also used. The capacity of the resins, before regeneration is needed, varies from 300 to 400 $m^8$ of raw water per $m^3$ of resin. During regeneration the plant has to shut down. Continuous ion exchange can be achieved only by using several plants.

- **ion pair**
a positive ion and an electron or negative ion produced by the action of ionising radiation.

- **ion selective electrodes**
electrodes that may be used to measure the concentration of ions in water, such as ammonium, nitrate, sulphide, chloride, potassium, etc. The presence of other elements may interfere with their accuracy and this may limit their use.

- **ionic bond**
chemical bond between two atoms involving the transfer of one or more electrons. Each transferred electron is equivalent to one ionic bond.

- **ionic equilibrium**
the condition in which the rate of dissociation of nonionised molecules is equal to the rate of combination of the ions.

- **ionic layer**
around a colloidal particle the nutrient elements seem to be distributed in two layers, in one they are negative, adhering to the particle (the inner layer which gives the particle its charge), in the other they are positive, looser, more or less mobile (outer layer, or of the opposite sign of that of the inner layer), these positive ions

are the exchangeable cations.

- **ionisation chamber**
an instrument for measuring radiation. It operates by measuring the ions produced by the radiation in a given volume and consists of two electrodes between which an electric field is maintained to collect the charge.

- **ionisation degree**
the proportion of potential ionisation that has taken place for an ionisable material in a solution or reaction mixture.

- **ionisation potential**
radiation capable of causing ionisation of the matter through which it passes. Ionising radiation may damage living tissue.

- **ionising event**
event in which an ion is produced.

- **ionising particle**
any charged particle possessing sufficient energy to produce ion pairs.

- **ionising radiation**
radiation that can deliver energy in a form capable of knocking electrons off atoms, turning them into ions.

- **iron bacteria**
organisms which are capable of utilising ferrous iron (either from the water or from steel pipe) in their metabolism and precipitating both ferric hydroxide in their sheaths and gelatinous deposits. These organisms tend to collect in pipe lines and tanks during periods of low flow, and to break loose in slugs of turbid water to create staining, taste, and odour problems.

- **iron chlorosis**
the yellowing of plants due to the lack of chlorophyll when the iron supply to plant is inadequate and its translocation to sites of physiological activity is retarted. Under certain situations, the chlorotic plants had as much and even more Fe than the comparable normal leaves. The inactivation of iron after it is absorbed accounted for chlorotic condition.

- **iron sulphate (copperas)**
is a granular material, which usually have a high degree of purity and is soluble in water. When applied to soils, it dissolves in the soil water and decomposes to form sulphuric acid, which in turn supplies soluble calcium through its reaction with lime. It can be used for reclaiming sodic soil as well as a source of iron for plants.

- **ironpan**
a thin compact layer within the soil consisting of an accumulation of sesquioxides, notably of iron. In humid climates, ironpans are characteristic of the B horizons of podzols and are formed by the redeposition of iron after leaching from above.

- **irradiation**
treatment with ray, as ultraviolet rays, X-rays, etc, the spreading of an effect of a stimulus, spreading of an excitatory process, apparent enlargement of objects, due to difference in illumination.

- **irrigable area**
the portion of the arable area of an irrigation development which is subject to irrigated farm use. It excludes lands required for non-productive uses.

- **irrigate**
to supply water to the soil for plant growth in place of, or in addition to, natural precipitation, using surface or underground sources of water.

- **irrigation**
the artificial application of water to the land in order to grow crops or to improve crop yields.

- **irrigation canal**
a canal constructed and used primarily for transporting water to be used for irrigating crops.

- **irrigation need**
is a measure of the amount and frequency of irrigation needed to improve the soil moisture deficit that develops when rain fails to improve the amounts of water removed by transpiration.

- **irrigation period**
the number of days that can be allowed for applying one irrigation to a given design area during the peak consumptive use period of the crop being irrigated. It is the basis for irrigation system capacity and equipment design. The irrigation system must be so designed that the irrigation period is not greater than the irrigation frequency.

- **irrigation requirement**
the total amount of water applied to the land surface in supplement to the water supply through rainfall and soil profile, to meet the water needs of crops for optimum growth. Also referred to as irrigation need or grant-in-aid.

- **irrigation structure**
any structure or device nec-

essary for the proper conveyance, control, measurement or application of irrigation water.

- **irrigation use of water**
the use of water primarily for the purpose of irrigating crops.

- **irrigation water**
water which is artificially applied in the processes of irrigation. It does not include precipitation.

- **irrigation wells**
a hole in the earth down to a supply of water excavated for bringing ground water to the surface.

- **isallobar**
a line on a map joining places where the same change in atmospheric pressure has taken place during a specific time. They are plotted to reveal to the forecaster how a pressure system is developing and moving.

- **isanomal line**
a line on a map joining places with an equal difference from the mean or normal of any climatic element.

- **isocitric acid**
an acid of the Krebs' cycle which is oxidised to oxalosuccinic acid, using NAD as a hydrogen acceptor, by the enzyme isocitric dehydrogenase.

- **isohumic coefficient**
the ratio Kj of the increase in the percentage of organic matter in the ground to the quantity of dry organic matter added in a given time by manures. It depends on the nature of the humus, varies from 0.10 to 0.50 It is expressed as per cent in dry soil.

- **isomorphous substitution**
the replacement of one ion by another of similar size in a crystal lattice without disrupting or changing the crystal structure of the mineral. The substituting ion may have a greater, equal or lower charge than the ion for which it substitutes.

- **isomotic solution**
a solution that has the same osmotic pressure as the cellular contents.

- **isopleth**
1. a line showing the variation

in time and position along a field profile of the point of interaction of a water-table contour line and the profile.
2. a graph showing, the occurrence or frequency of any phenomenon as a function of two variables.

- **isoquant**
1. it is the locus of all combinations of Xi and X> to produce a given level of output. It is usually expressed as—Xi=g (Xj, Y°). Where Y° is a specified level of output.
2. it refers to a curve which shows the various possible combinations of productive factors yielding a given quantity of output.

- **isothermal process**
a process in which the temperature remains constant. In such a process, the system is in equilibrium with its surrounding throughout.

- **isotope effect**
the slight difference in biological properties and metabolic fate of labelled and unlabelled compounds because of differences in chemical behaviour of different isotopes of the same element.

- **jacquemart's reagent**
analytical regent used for testing ethylalcohol, consists of an aqueous solution of mercuric nitrate and nitric acid.

- **joint products**
two or more goods produced by a single process in such a manner that it is impossible to produce one good without turning out the other. Such goods are produced under conditions of joint cost and joint supply. Examples of joint products include wool and mutton, beef and hide.

- **jointed**
said of an elongated plant member which is constricted at intervals, and ultimately separates into a number of portions by breaking across the constrictions.

- **jointing stage**
the stage when grass stems begin elongating.

- **juglar cycle**
it means a business cycle of intermediate duration. It is named after Clement Juglar who first discovered it. Juglar cycle refers to fluctuations of prices, production, employment, etc., over a period of nine to ten years.

- **juvenile form**
a young plant that has leaves and other features different

from those of a mature plant of the same species.

- **juvenile leaf**
the form of leaf found on a sporeling or seedling, when it differs markedly from the leaf of the adult plant.

- **juvenile stage**
a special stage in the life history of some algae from which the ordinary plant develops as an outgrowth.

- **juvenility**
in some plants, the difference in the interval stages of young seedlings from that of mature plant which is manifested by striking differences in leaf shape or greater thorniness in younger plants. This immature condition is referred to as juvenility. It is important in several aspects of plant propagation.

- **juxta**
a combining form meaning nearness or situated near.

- **k capture**
a term used for 'K-electron capture' or 'electron capture'.

- **kainit**
a potash salt containing potassium and sodium chlorides and sometimes sulphate of magnesia, with not less than 12 per cent potash (KzO).

- **kaipad system of paddy cultivation**
this is a method followed in Kerala for growing paddy in lands which are normally flooded by saline water. Number of small mounds of 2-3 feet height are prepared in the field during March and April each year. Paddy is sown when the early monsoon showers of May wash away the salts in the top layers of the soil, to some extent. After sufficient growth of seedlings, the mounds are cut into smaller bits and the seedlings are planted when the monsoon is active.

- **kame**
an irregular ridge or hill of stratified glacial drift.

- **katteklel**
a Dutch term, Iron sulphide containing material formed from brackish-water clay as the result of the oxidation of iron sulphate which gave acid

and iron sulphate when there was not sufficient lime to neutralise the acid that had formed.

- **keel**
  the two small petals opposite the banner or largest petal of the legume flower, in some species they are fused together.

- **kelp**
  an aquatic plant from which methane can be produced.

- **kelvin absolute temperature scale**
  a temperature scale in which the ratio of the temperature of two reservoirs is equal to the ratio of the amount of heat absorbed from one of them by a heat engine operating in a Carnot cycle to the amount of heat rejected by this engine to the other reservoir, the temperature of the triple point of water is defined as 273.16 K. Also known as Kelvin temperature scale.

- **kelvin equation**
  an equation giving the increase in vapour pressure of a substance which accompanies an increase in curvature of its surface.

- **key industry**
  an industry which is of importance to other industries. Steel, for example, is a key-industry.

- **keynes' effect theory**
  a reduction in the general price-level results in lower interest rates and increased investment.

- **khaira disease**
  a zinc deficiency disease of rice observed in Tarai region. About two weeks after transplanting, the disease starts appearing in low lying areas in more or less circular patches in the field. The disease is characterised by the appearance of brownish-red (rusty) discolouration on the surface of outer leaves. The tips of many leaves become dry and yellowing followed by rusty discolouration proceeds downwards from the tips along the margin of the leaves. Plant growth is stunted.

- **kharif season**
  the season in which crops are grown by making direct use of the rainfall.

- **khurpi**
  a multipurpose hand tool used for interculturing and weeding on the farm. Rural blacksmiths make it.

- **kiln**
  an oven, furnace or heated

enclosure used for processing a substance by burning, drying or firing.

- **kilo**
a prefix denoting $10^3$. For example, 1 kilometre (km) = $10^s$ metres (m).

- **kilocurie**
one thousand curies.

- **kilowatt hour (kwh)**
practical unit of work for energy. It is equal to the energy supplied by one kilowatt of power in one hour. I kwh=3.6x $10^6$ joule.

- **kin selection**
the evolution of characteristics which favour the survival of close relatives of an individual.

- **kinetic energy (t)**
the energy possessed by a body by virtue of its being in motion. For a body of mass (m) moving with velocity V, the kinetic energy is given by $T=|m V^2$.

- **klelplant soils**
soils consisting of heavy clay, but with a peat layer, lying so deep and being so thin that the area did not go inversion, *i.e.* sequence of clay and peat layers was not reversed.

- **klendusity**
the ability of a susceptible plant to escape infection under conditions in which other susceptible varieties contract the disease.

- **klimamorphologie**
the science of the study of the relationship between climate and morphological features.

- **koeppen's classification, climate**
classification scheme which relates climate to vegetation while providing an objective, numerical basis for defining climate types in term of climatic elements.

- **kraal method of composting**
a method of composting in which bedding and other plant residues like straw, maize stalks and leaves, bed silage, grass hay or grass specially cut for the purpose and weeds are carted into Kraals, and packed in layers of approximately 40 to 60 cm high and left there until thoroughly mixed with dung and urine of the animals. This method can be adopted in areas with good rainfall in farms where there is a considerable number of livestock.

- **kreb's cycle**
a cycle sequence of reactions which plays a central role in

the metabolism of many heterotrophic microorganisms capable of respiratory metabolism. It involves an enzyme which converts pyruvic acid to carbon dioxide in the presence of oxygen with concomitant release of energy which is captured in the form of ATP molecules. During each complete turn of the cycle one molecule of acetyl-CoA enters the cycle, two molecules of carbon dioxide are eliminated, three molecules of NAD (P) and one of FAD are reduced, and one molecule of oxaloacetate is regenerated. The reduced enzymes ADP, fad are reoxidised via the electron transport chain with concomitant release of energy in the form useful to the cell. The Kreb cycle is also referred to as the citric acid cycle or the tricarboxylic acid (tca) cycle.

- **krotovinas**

are irregular tubular streaks within one horizon of material transported from another horizon. They are caused by the filling of tunnels made by burrowing animals in one horizon with material from outside the horizon.

- **kurtosis**

the departure of a symmetrical frequency distribution from the normal by excess (platykurtosis) or deficiency (leptokurtosis) in its shoulders, as opposed to tail and centre.

- **kutter's formula**

an empirical formula expressing the value of the coefficient, C, in the chezy formula, in terms of the friction slope, hydraulic radius, and a coefficient of roughness.

- **kwelders**

soils along the coast of Holland subject to overflow by sea water at high tide.

- **labelled compound**

a compound in which a stable atom is replaced by a radioactive isotope of that atom. The path taken through a mechanical or biological system by such a labelled compound can be traced by the radiation emitted by the labelled atom.

- **labelling**

all information and other written, printed or graphic matter upon the agricultural chemicals and seeds or any of its accompanying containers or wrappers to which reference is made on the label or in supplemental literature accompanying the economic poison.

- **labile chemical bonding**
  the temporary holding of cations and anions in more labile complexes.

- **lacustrine peat**
  the decomposition of vegetation to form peat, under anaerobic conditions.

- **lacustrine soil**
  soil formed at the bottom of a lake by deposition of mineral particles removed from banks or carried by flowing water, or of debirs of organisms.

- **lag**
  the time that elapses between a change in a condition or happening of some event, and the recording of the same by an instrument.

- **lag period**
  the period between the initial inoculation of a culture medium with an organism and the resumption of normal growth.

- **lag phase**
  the 1st phase of bacterial multiplication, in which there is no appreciable reproduction of the bacteria takes place.

- **lamarckism**
  the evolution theory of Lamarck, embodying the principle that acquired characteristics are inherited.

- **laminated**
  refers to a soil arranged in very thin layers or plates, horizontal to the soil surface.

- **land**
  the indestructible power of soil is called land. This refers to agricultural land and includes the fertility of soil, air, water and sunshine as part of it.

- **land grading**
  is reshaping of the field surface to a planned grade. It is necessary in making a suitable field surface to control the flow of water, to check soil erosion and provide for surface drainage.

- **layer (litter)**
  the surface layer of the forest floor consisting of freshly fallen leaves needles, twigs, bark, stem and fruits. The d horizon.

- **light requirement**
  the amount of light necessary for the satisfactory development of a species.

- **light seed**
  a seed which will not germinate unless it has been exposed to light.

- **ligno humus complex**
a dark reddish substance dispersed in the soil, intermediate between humus and lignin. Soluble in alkalis and water and is destroyed by acetyl bromide. Belongs to fulvic acid group. Ligno humus complex usually accumulates in the B horizon.

- **lime coating or pelleting**
applying a coat or coats of very finely ground limestone or similar material to protect the inoculant.

- **lime concretion**
an aggregate cemented by precipitation of $CaCO_3$.

- **lime elements**
the essential primary elements, calcium and magnesium. They are so called because they are usually applied as found in lime.

- **lime pan**
an aggregate cemented by precipitation of calcium carbonate.

- **lime requirement**
the number of Kilograms of limestone or other specified liming material required to raise the pH of one hectare furrow slice (2,000,000 kg) of an acid soil to any desired value under field conditions.

- **lime sulphur (calcium polysulphide)**
is a yellowish-brown, alkaline liquid, which usually contains about 24 per cent of sulphur. The calcium content is about one-fourth that of sulphur. The action of lime sulphur in soils depends mainly on the sulphur content. Like elemental sulphur, it must first be oxidised to sulphuric acid, and then react with lime to produce a soluble form of calcium.

- **limestone (caco$_3$)**
a sedimentary rock, largely calcium carbonate, usually also containing significant amounts of magnesium carbonate. It is practically insoluble in alkaline soils but it goes into solution slowly in acid soils. The rate at which it goes into solution and supplies soluble calcium depends upon the degree of fineness. Waste lime from sugar factories depends upon the degree of fineness. Waste lime from sugar factories is a desirable source of calcium carbonate for application to soils because the particles are fine. Limestone is sometimes applied to

lime free soils, along with acid or acid-forming amendments.

- **liming**
  the application of lime to land, primarily to reduce soil acidity and supply calcium for plant growth. Dolomitic limestone supplies both calcium and magnesium. May also improve soil structure, organic matter content, and nitrogen content of the soil by encouraging the growth of legumes and soil microorganisms. Liming an acid soil to a pH value of about 6.5 is desirable for maintaining a high degree of availability of most of the nutrient elements required by plants.

- **liming factor**
  1. the quantity of calcium carbonate necessary to raise the pH by 0.1 for one hectare of surface soil to a depth of 10 cm. 2. the coefficient by which it is necessary to multiply the quantity of lime necessary for neutralisation in the laboratory, to obtain neutralisation in the soil in place. The value of this factor ranges from 2-3 due to soil heterogeneity.

- **liming soil**
  application of ground lime stone or slaked lime to the soil to lower the acidity of the soil. This is generally practised in temperate zones.

Liming on the Big Island

- **limit of resolution, resolving power**
  the size of the smallest object that can be seen through a microscope, telescope, etc. In the optical microscope, it is about 0.25 *fan* or 250 mm, in the electron microscope about 0.1 mm. Some microbes are more difficult to see than others.

- **limiting factor**
  abiotic condition that most controls the growth of a species. For most terrestrial plants this condition is the supply of the nutrient nitrogen in the soil.

- **line breeding**
  the mating in successive generations of individuals having the same common ancestor.

- **lineage**
  the kinship relation between an individual and the

individual's progenitors.

- **linear flow**
flow in which the streams are rectilinear and parallel.

- **linear programming**
a mathematical method of calculating the least-cost combination of the factors of production when there are many possible alternative combinations available to the firm.

- **linear regression**
the regression represented by a straight line is spoken of as the linear regression.

- **lint length (cotton)**
the length of unicellular spinnable hairs springing from the seed coat, collectively known as lint, the length varies from less than 1/8 of an inch to *1* inches.

- **linters**
the short fibers that remain on cottonseed after ginning. They are used mainly for batting, mattress stuffing, and as a source of cellulose.

- **lipid hypothesis**
the concept that solubility in a lipid is important in determining the rate at which substances are transferred across membranes.

- **lipoid solubility hypothesis**
according to Overton (1897) the uptake of salts depended on their relative solubility in lipoid which was a constituent of the cell membrane.

- **lipophilic**
molecules or substances attracted by oil-like surfaces and molecules but repelled by water.

- **lipophobic**
molecules or surfaces attracted towards water but repelled by oil-like surfaces or substances.

- **liquid farm manure**
solid and liquid manure from hog and dairy operations having little bedding on concrete floors collected in storage tank with the help of 10-25 per cent water. This solid liquid slurry is agitated every 3-4 weeks and then pumped into a spreader tank. It could then be mixed with the soil. It saves labour besides preserving plant nutrients.

- **liquid fertiliser**
a fertiliser which would be in liquid form when stored under high pressure. Anhydrous ammonia is one such fertiliser. This can be applied direct to the soil in its liquid

form.

- **liquid waste**
dissolved and suspended solids in water in dilute concentrations generally not biologically stable, also refers to used or discarded non-aqueous fluids.

- **lithic contact**
a diagnostic contact to non-soil material. This is a boundary between soil and continuous coherent, underlying material that has a hardness >3 Moh.

- **lithochromic soil**
a soil which owes its colour to parent material or rock from which it was derived.

- **lithophyte**
a plant which likes stone. Devoid of an absorbent apparatus. Uses the substances dissolved in rain water and dew covering stones, *e.g.* algae, fungi, gelatinous lichens, etc.

- **lithosequence**
plant that grows on rocks or stony soil and derives nourishment from the atmosphere. Expressed mathematically as shown below:
S=f (pm) cl, o, r, t,
Where S=soil, f== function of, pm=parent material, Cl=cli-mate, o=organisms, r=relief, t-=age.

- **lithosols**
an azonal group of soils having no clearly expressed soil morphology and consisting of a mass of rock fragments from consolidated rocks which are imperfectly weathered. They are found primarily on seteeply sloping land .

- **lithotroph**
organism that uses an inorganic substrate such as ammonia or hydrogen as an electron donor in energy metabolism. There are two types of lithotrophs: chemolithotroph and photolithotroph.

- **litmus paper**
unsized paper treated with litmus for use as an acid-base indicator. Will turn red in acids and be restored to blue by an alkali. May also have red colouring matter that turns blue in the presence of an alkali.

- **litter**
the freshly fallen or only slightly decomposed plant material on the forest floor. This layer includes foliage, bark fragments, twigs, flowers, and fruit.

- **living mulch**
a cropping system in which the sod grows between the rows of the cultivated crop

such as corn. The corn may be grown in wider rows with thicker planting in the row and the legume or grass-legume mixture sown after the first or second cultivation.

- **livingston's cone**
a device for measuring the moisture of the soil from the weight, increase in a porous cone which has been sunk vertically into the wet ground.

- **lixiviate**
separate into soluble and insoluble components by percolation of water.

- **lixiviated soil**
leached soil, lixiviation natural process in the formation of soil, particularly in humid regions.

- **loamy**
a broad grouping of texture classes, includes all sandy loams, clay loams, loam, silt, and silt-loam textures.

- **loamy sands**
soil material that contains at the upper limit 85 to 90 per cent sand and silt plus clay, at the lower limit it contains not less than 70 to 85 per cent sand, and the percentage of silt plus twice the percentage of clay does not exceed 30.

- **local**
this term refers to the site of action of a poison and means that the action takes place at the point of area of contact. The site may be skin or mucous membrane.

- **local climate**
the climate of a small area which possesses marked contrasts with other areas nearby, resulting from minor differences of slope and aspect, colour and texture of soil, proximity of a water-surface, nature of vegetation cover.

- **lodging**
grass or grain in fallen or laid down condition due to fierce wind or heavy rain or inherent weakness of the plant.

- **loess**
a soil made up of small particles that were transported by the wind to their present location.

- **logarithmic growth**
a growth phase of bacteria (or other life) that involves very rapid increase in the numbers of a population. For bacteria, it means that there is no limitation of food. If a bacterium reproduces itself every 30 min and its successors do so, the growth will be logarithmic and after a day there will be 2.8 x $10^{14}$ or 280 million which have grown from one bacterium.

- **logarithmic phase**
the 2nd stage in bacterial multiplication and the most rapid, with the increase tending to occur in geometric progression.

- **logarithmic scale**
a scale of measurement in which an increase of one unit represents a tenfold increase in the quantity measured.

- **logarithmic sprayer**
a device for spraying a pesticide at a steadily decreasing rate of application.

- **logding**
is the inelastic displacement of the plant from the vertical or initial position. Usually wind provides the force that produces the lateral displacement. The moment of force, however, is determined not only by wind pressure but also by the height and weight of the plant and the weight of rain that may be adhering to the plant. Nitrogen fertilisation frequently increases plant height, weight and leaf area, and hence increases the moment of force.

- **logging**
drawing behind a tractor, a heavy log to flatten vegetation.

- **long day plant**
a plant which flowers after a critical day length is exceeded. The so-called critical day length differs from species to species. Some examples of long-day plants are *Spinacea oleracea* (Spinach), *Beta vulgaris* (sugar beet), and *Hyoscyamus niger* (Black henbane).

- **loosening**
the decreasing of soil bulk density and increasing porosity due to the application of mechanical forces to the soil.

- **low volume spray**
a spray application of 0.5 to 5.0 gallons per acre.

- **low volume spraying**
spraying, when spray material deposited in small drops do not form pools. Obtained by spraying liquid up to 20

gallon per acre. Saves labour, active ingredient and time.

- **lower critical limit**
that concentration of a nutrient element in plant or soil below which the plant can not complete its life-cycle.

- **lower plastic limit**
is that moisture content at which the soil can be rolled out into a wire. It represents the moisture content necessary for the change from friable to the plastic consistency. At this point sufficient moisture has been added to provide a film around each soil particle whereby sliding over each other occurs. Cohesion is maximum at the lower plastic limit.

- **lowland**
land lower in elevation than surrounding or adjacent land, and whose soil is often poorly drained.

- **lumpy structure**
soil in lumps, the irregular form and the volume of which depend on the way in which the soil has been broken up by implements.

- **lustre (mineral)**
the visual property of something that shines with reflected light.

- **lysis**
rupture of a cell, resulting in loss of cell contents. Under normal conditions such cells are engulfed by phagocytes and degraded by their lysosomes. Only under rare conditions are such cells degraded from within by lysosomes.

- **lysozyme**
an enzyme present in some plants that hydrolyses glycosidic linkages within peptidoglycans in prokaryote cell walls. Present in some plantslytic pertains to the dissolving of a substance as the result of the action of another substance *e.g.* the dissolving of certain bacteria by lysozyme.

- **M**
a symbol indicating strong cementation, induration. It is applied as a suffix to horizon designations to indicate irreversible cementation. The symbol is not applied to indurated bedrock. Contrary to previous usage, m is not used to indicate firmness, as in fragipans, but is confined to indurated horizons-which are essentially (more than 90 percent) continuous, though they may be fractured.

- **macro**
a prefix meaning large.

**maculates**
Latin meaning 'spotted'.

**made land**
consists of areas filled artificially with earth, trash, or both, and smoothed. It occurs most commonly in and around urban areas. Stabilised land areas with clearly developed soil characteristics or even those with young soils if definable and uniform enough to map, and especially if arable, should be classified as soils even though originally made or reworked by man.

**mafi**
temporary tenancy of land provided in exchange for services rendered. This is prevalent in Uttar Pradesh (India).

**magatha**
a system of engaging farm labour on the basis of share-cropping on an oral lease for a period of a year or two. This system is prevalent in Andhra Pradesh (India).

**magic numbers**
a nucleus is more stable if its neutron and proton numbers are 2, 8, 20, 28, 50, 82 or 126 called the magic numbers.

**magmatic water**
water arising from an underground magmatic source.

**magnesia (magnesium oxide)**
a product consisting chiefly of the oxide of magnesium. Its grade is stipulated, for example, Magnesia—75 per cent MgO.

**magnesia mixture**
it is used for distinguishing arsenates from arsenites. It forms white precipitate with arsenate. No precipitate is formed with arsenites. Magnesia mixture contains 110 gm of magnesium chloride dissolved in 700 ml of ammonia and diluted to 2 litres.

**maintenance behaviour**
behaviour which allows an animal to carry out its day-to-day activities such as search for food, mating, reproduction, or avoidance of extreme environments.

**maintenance ration**
the ration which helps maintain the general level of health of the animals and enables them to carry on their normal functions.

**mallee soil**
brown, highly calcareous soil of the semiarid regions of south eastern Australia. The subsoil containing more clay

and calcium carbonate than the surface soil is typically strongly alkaline and rather saline.

- **maltose**
a sugar molecule consisting of two linked glucose molecules. Maltose is the most common type of sugar in beer wort, typically comprising about 50% of the sugars present.

- **marketing machinery**
a synonym for the food marketing system.

- **marketing margin**
the portion of the consumers' food money paid to food marketing firms for their services and value-adding activities, the 'price' of all food marketing activities.

- **marketing plan farm**
a set of objectives, strategies, and tactics which guide a farmer's production and marketing decisions.

- **marketing strategy**
a procedure used to achieve a market goal, for example, one grocery store may use the strategy of low process to attract consumer, another might employ the strategy of high quality.

- **markup price**
the price of a good after some percentage allowance for profit has been added to the cost price.

- **marl**
an unconsolidated mineral deposited in marine or fresh water and consisting chiefly of silt- and clay-sized particles of calcium carbonate.

- **marl beds**
include areas where marl (an earthy crumbling deposit consisting chiefly of calcium carbonate mixed with clay or other impurities) is exposed or lies a few inches below the surface. Marl is usually formed in lakes or ponds. The calcium carbonate may have originated from the calcareous remains of the chara plant (chara marl), from mollusc shells (shell marl), or from simple precipitation from solution. These beds are commonly a good source of agricultural lime.

- **mass curve**
a graph of the cumulative values of a hydrologic quantity (such as precipitation or runoff), generally as ordinate, plotted against time or date.

**mass defect**
the sum of the masses of the various constituents of the atom is always found to be more than the atomic weight of the atom. This difference is called mass defect as shown below:
mass defect=Znu+(A—Z)m«—M.
where Z is the atomic number, ma and $m_n$ represent the masses of the hydrogen atom and the neutron, respectively, A—Z is the number of neutrons in the nucleus and M is the atomic weight.

**mass flow**
hypothesis of transport of materials in phloem. Sieve tubes and contents are regarded as forming an osmotic system continuous throughout the plant connected, via plasmodesmata with other tissues. A gradient of osmotic and hence turgor pressure exist within the sieve-tube system as a result of entry into it of synthesised materials. This pressure gradient results in mass flow of water and dissolved substances through the system from regions of high to regions of low pressure.

**mass market**
popular market for mass produced goods as distinct from custom built, made to measure, and other individual products.

**mass production**
repetitive machine production of large quantities of identical products.

**mass spectrometer**
an apparatus for obtaining the mass spectrum of a beam of ions by means of suitably disposed magnetic and electric fields. The deflection of any individual ion in a magnetic field depends on the ratio of its mass to its electric charges, m/e. Typically, a mass spectrum is obtained by deflecting the ions onto a thin slit and detecting the ion current with an electrometer. The magnetic field is varied so that different types of ion are detected at different field values.

**mass spectrometry**
an analytical technique for identification of chemical structures, determination of mixtures, and quantitative elemental analysis, based on application of the mass spectrometer.

**masscuite**
a sugar factory term used for mixture of sugar crystals and molasses.

**massive**
a structureless compacted condition of soil, showing no distinct arrangement of soil particles.

**master factor**
any powerfully acting ecological factor, which plays the main part in determining the occurrence in a given area of a plant community of major rank.

**maternal inheritance**
the occurrence of heritable differences referable to materials transmitted by the egg, but not through the male gamete.

**matrix effects**
the enhancement or suppression of minor element spectral lines from metallic oxides during emission spectroscopy by the matrix element (such as graphite) used to hold the sample.

**maturation**
ripening, completion of germ-cell development.

**mature green stage**
1. stage at which tomato fruit on plant is still green, but fully developed, and at which the pulp surrounding the seed in the fruit cavities is jelly-like and seeds give way to the edge of a sharp knife and are not cut into slices.
2. ready to harvest for distant markets.

**mature soil**
is one that has assumed the profile features characteristic of the predominant soils on the smooth uplands within the general climatic and botanic region in which it is found.

**maximum and minimum thermometer**
a thermometer designed to record the maximum and minimum temperatures during a period of time. It usually consists of a graduated capillary tube at the base a bulb is found containing ethanol. Within the capillary is a thin thread of mercury with a steel index at each end. When the temperature rises, the alcohol expands, thus forcing the mercury up the tube, carrying the upper steel index. Therefore the position of the upper index marks the maximum temperature reached, and similarly the position of the lower index, the minimum temperature attained.

**maximum available water**
1. the quantity of water represented as the difference between field capacity and wilt-

ing coefficient.

2. This is the moisture that can be readily extracted by the plant and used for normal growth.

- **maximum computed flood**

the largest momentary flood discharge believed possible from a consideration of meteorological conditions and snow cover on the watershed.

- **maximum dosage**

the largest amount of a pesticide that is safe to use without resulting in excess residues or damage to whatever is being protected.

- **maximum known flood**

the greatest flood which has occurred within the memory of inhabitants of a region.

- **maximum likelihood method**

a method of estimating a parameter(s) of a population by that value(s) which maximises the likelihood of a sample.

- **maximum moisture holding capacity**

1. the amount of moisture in a soil when its total pore space—both macro and capillary—is completely filled with water. Under these conditions gravity does not come into play and the soil moisture tension is very low between l/100th to l/1000th of an atmosphere or pF=0 to 1. No crops, except a few crops like paddy, can stand such saturated conditions devoid of aeration, for long periods of time.

2. Maximum moisture holding capacity is also known as maximum retentive capacity.

- **maximum permissible dose**

for an individual, the term refers to the dose that accumulated over a long period of time or results from a single exposure, which in the light of present day knowledge carries a negligible possibility of severe somatic organic injuries. Furthermore, it is such a dose that any effects that ensure more frequently or limited to those of a minor nature that would not be considered unacceptable by the exposed individual and by competent medical authorities.

- **maximum probable flood**

the maximum flood for which there is a reasonable change that it will occur on a given stream at a selected site. It is often assumed to be equal to maximum flood observed in areas having the same or similar physiological and meteorological characteristics.

- **maximum probable precipitation**
precipitation of a given amount and duration that can reasonably be expected to occur in a drainage basin.

- **maximum proctor compaction**
through which the density of the soil passes when the soil is subjected to a given type of compaction, when the moisture content is increased starting from zero.

- **maximum stream flow**
the maximum rate of discharge during a specified period.

- **maximum sustained annual yield**
term used in forestry for securing to the fullest possible extent a sustained annual or periodic return from a forest.

- **mciotherm**
a plant that thrives in a cool temperate environment.

- **meadow**
an area covered with fine-stemmed forage plants, wholly or mainly perennial and used to produce hay.

- **meadow soil**
1. an intrazonal hydromorphic soil that occurs in low-lying areas such as flood plains. It has a humus-rich dark brown
2. a-horizon that overlies a mottled gleyed zone.
3. lush grassland, often managed for a hay crop, provides the surface humus.

- **meadow strip**
a sloping field or strip of grassed land which, in addition to yielding a hay crop, acts as a broad shallow water channel during period of run-off. Often used as a terrace outlet channel and usually much larger in extent than a grassed waterway.

- **mean**
the sum of any series of numerical items as such sources or observations, divided by the number of items in the series.

- **mean annual increment**
the total growth of a stand of trees divided by the total age.

- **mean cycle length**
the average of the interval between any two consecutive heats of animals, when longer

than normal, indicates embryonic death with the presence of venereal disease or undetected heat.

- **mean deviation**
the arithmetic average of deviations of all the values, taken from a statistical average (mean, mode or median) of the series. In taking deviations of values algebraic signs + and — are not taken into considerations, that is negative derivations are also treated as positive deviations.

- **mean free path**
average distance travelled by a particle between collisions.

- **measurable precipitation**
any precipitation equivalent to over 0.005 in. of rainfall. Less than this considered as trace.

- **measure weight test**
a test which gives an indication of the relative heaviness, pulpiness and quality of seeds. For this purpose, a specified volume of seed is weighted and represented as kilos per hecto litre.

- **measuring weir**
a notch of regular form, usually sharp-crested, used for measuring the flow of water.

- **meat meal**
is the material obtained by either steaming carcasses of animals at a pressure of 50 to 60 lbs for a few hours or by cooking them in specially improvised pans. The process of cooking under steam pressure and subsequent drying completely make this manure safe for handling. During steam-heating, the bones get brittle, and it is easy to powder them. It contains 8—12% N and 0.1—3.5% $P_2O_5$.

- **mechanical analysis**
the separation by mechanical means of the different size groups (separates) and the determination of the percentage of each group in a given soil sample (or a particle-size analysis).

- **mechanical control**
the control of erosion through structures, such as dams, terraces, etc. in contrast to vegetative control.

- **mechanical flocculation**
a low-cost method of flocculation capable of improving the clarity of raw water or raw sewage as well as of effluents, by the gentle movement of a picket-fence stirrer or of paddles at speeds of the order of 70 mm/s for 15 min,

followed by 30 min of quiescence. It is often used after coagulation.

- **mechanical register**
an electrically actuated mechanical device for recording counts or impulses.

- **mechanical scarification**
scarification is any process of breaking, scratching, or mechanically altering the seed covering to make it permeable to water or gases. Sacrification should not proceed to the point at which the seeds are injured. Although some sacrification probably occurs during harvesting, threshing, and cleaning, germination of most hard-coated seeds is improved by additional artificial treatment.

- **mechanical seed drill**
machine which automatically drops seeds in lines, the feeding being externally forced, internally forced, or having spoon-feed for distributing the seed. Power transmitted to seed distributing wheel by gears attached to ground wheel moving as the bullocks pull the drill. Seed placed in seed-box, and picked up by cups or notches and dropped in funnel to be transmitted by tubes to the furrows.

- **mechanical ventilation**
process of supplying or removing air with a mechanical device to or from any space, such as interstitial space between products, used in connection with conditioning of harvested produce.

- **mechanism**
a step-by-step description of the events taking place in a chemical or biochemical reaction.

- **megatherm**
a tropical plant, a plant requiring moist heat.

- **megistotherm**
a plant that thrives at a more or less uniformly high temperature.

- **meiophase**
the stage during which meiosis occurs so that the chromosome number is reduced from diploid to haploid.

- **meiosis**
the reduction division, ultimately resulting in the production of gametes. Two suc-

cessive divisions halve the chromosome number. The nucleus divides twice, but the chromosomes only once.

- **melanisation**
  the darkening of light-coloured mineral initial unconsolidated materials by admixture of organic matter.
- **mellow soil**
  a very soft, very friable, porous soil without any tendency towards hardness or harshness.
- **melting point**
  a characteristic temperature at which the solid state of a substance can stay in equilibrium with its liquid state under a pressure of one atmosphere.
- **membrane**
  a thin film, skin, or layer of tissue covering a part of an animal or plant, a thin covering of cells or of unicellular organism.
- **mendelian population**
  a group of interbreeding individuals which share a common gene pool, the species being the most extensive.
- **mendelism**
  a law or rule governing inheritance of characters in plants and animals, discovered by Gregor Mendel. This principle deals with inheritance of 'unit characters' presence or absence of one or other of a pair of contrasting characters, dominant and recessive. It also shows that offspring of organisms with a pair of contrasting characters will exhibit these in a definite ratio, and it is extended to deal with groups of characters.
- **meristematic ring**
  tube of maristematic tissue between cortex and pith, subtending the apical meristem and giving rise to vascular tissues.
- **meromixis**
  a type of genetic exchange in bacteria where transfer of genetic material is in one direction only.
- **mesh sizes**
  the mesh numbers used in the past to describe sieve openings indicated the number of holes (or wires) per lineal inch, thus, a higher number had smaller openings. The diameter of the wires affects the hole size. Sieves therefore are now defined by the size of opening and the wire diameter in mm, under BS 481.
- **mesic and isomesic**
  one of the soil temperature categories. 8°Cto <15°(47°to59°F).

- **mesobiota**
soil organisms ranging in size from those just visible with a hand lens to those several centimetres long, *i.e.* the middle size range.

- **mesoclimate**
1. the climate of small areas of the earth's surface that may not be representative of the general climate of the region.
2. The mesoclimate is intermediate in scale between macroclimate and microclimate.

- **mesoderm**
the embryonic layer lying between ectoderm and endoderm.

- **mesohydrophytic**
growing in temperate regions but requiring much moisture.

- **mesomitosis**
mitosis which take place within the nuclear membrane, without any cooperation from cytoplasm elements.

- **mesophilic bacteria**
bacteria which grows best at a temperature of 10-40°C.

- **mesophyll resistance**
the resistance offered to gaseous diffusion by the intercellular spaces of leaves.

- **mesophytic environment**
a environment in which the water-supply is neither very scanty nor abundant.

- **mesoplasm**
the central zone of the cytoplasm.

- **metabolic pathway**
a set of consecutive intercellular enzymatic reactions that converts one molecule to another.

- **metabolic stage**
the resting stage between two successive cell divisions.

- **metabolism**
the sum of all the physical and chemical processes by which living organised substance is produced and maintained and by which energy is made available for the uses of the organism.

- **metabolite**
any substance produced by metabolism.

- **metamitosis**
mitosis in which the nuclear membrane disappears and the karyokinetic figure lies free in the cytoplasm.

- **metamorphic rock**
a rock which constitution has undergone pronounced alteration. Such changes are gen-

erally effected by the combined action of pressure, heat, and water, frequently resulting in a more compact and more highly crystalline condition of the rock. Gneiss, schist, and marble are common examples.

- **metaphase**
  the stage of cell division in which the chromosomes are arranged in an equatorial plate or plane, precedes the anaphase stage.

- **metaphase plate**
  the group of chromosomes arranged in the equatorial plane of the spindle.

- **metaplasm**
  any substance within the body of a cell which is not protoplasm, especially food material.

- **metastable state**
  an excited-energy state of a nucleus which returns to the ground state by the emission of a gamma ray over a measurable half-life.

- **metastructure**
  a soil structure formed due to the surfaces of fracturation (large-scale structure).

- **micro fauna**
  the comprise protozoa, nematodes and similar organisms. These form part of the lowest rung of the animal kingdom and are so small that they cannot be distinguished without the help of a microscope.

- **micro ph**
  variations of the pH within a distance of a few millimetres. In the soil the pH values of neighbouring particles can perceptibly different *e.g.* from 6.5 to 8.5 for a soil of pH 7.5. It is measured using Wulff's method with small strips of cellulose measuring 0.5 mm.

- **microanalysis**
  qualitative or quantitative analysis of chemicals that are in quantities of the order of grams.

- **microbial insecticide**
  pathogen used for killing insects. Free from harmful residues, non-phytotoxic to crops of higher doses, harmless to beneficial insects.

- **microbial processes**
  the activities and influence of microorganisms in modifying soils or other substances.

- **microeconomics**
  is that branch of economic analysis which studies the economic behaviour of the individual unit, may be a person, a particular household or a particular firm. It is a study

of one particular unit rather than all the units combined together. In microeconomics we examine the trees, not the forest.

- **microenvironment**
the environment close enough to the surface of living or nonliving object to be influenced by it.

- **microerosion of the soil**
erosion on the microscopic scale.

- **microfibril**
it is believed to have been formed from about 20 micelles and has a cross-sectional area of about 62500 A. Micro-fibrils are seen under electron microscope.

- **microflora**
1. living or fossil animals too small to be seen with the naked eye. The term is sometimes used synonymously with paleoflora but is more frequently applied to living microscopic algae and fungi. 2. a very localised or small group of plants, occupying a very small habitat.

- **microgram**
a metric weight measurement equal to 1/1,000,000th of a gram, approximately 28,500,000 micrograms equal to 1 ounce.

- **microhabitat**
the particular parts of the habitat that an individual encounters in the course of its activities.

- **micromanipulator**
a complicated piece of apparatus used for fine dissection under the microscope, or for single cell isolation.

- **micrometeorology**
the science concerned with the detailed examination on a micro scale of the physical and meteorological processes taking place within a small distance, for example, between the top of a plant to its roots.

- **micrometer**
the unit of microscopic measurement, 1/1,000 of a millimetre, represented by ^m.

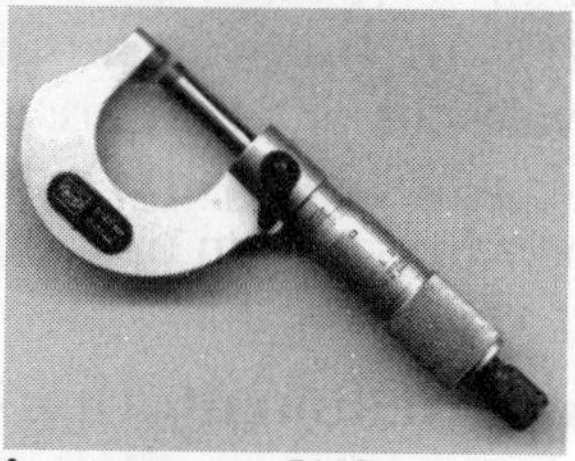

- **micromonolith**
a small soil sample taken with a knife, which can be fixed and glued on cardboard for demonstration or study.

- **micronutrient fritted glass**
fritted trace elements are those

incorporated into a silicate or glass matrix to form a powdered material called frits. The powdered raw materials are dried and mixed in batches, fused in a smelted furnace, and this molten matrix quenched, dried and milled. It is then compacted and granulated before bagging and storage. Solubility of metallic salts in the frits is controlled by particle size and changes in the composition of the matrix. More than one micronutrient may be included to provide mixes for various crops and soils.

- **micropore**
  a very small pore within a lump, from 0 to 10 microns those of 1 to 10 microns can harbour bacteria.

- **microrelief**
  small-scale differences in relief, including mounds, swales, or pits that are a few feet across and have differences in elevation up to about 6 feet.

- **microsedimentation**
  sedimentation on the microscopic scale. In the voids of the soil, the matter torn from the walls, dissolved or peptised, is deposited at other points in the form of coatings, tubes, etc. It sometimes completely obstructs the cavities.

- **microsol**
  small, narrowly localised strips of ground.

- **microstructure**
  1. structure on an atomic scale.
  2. applied to soil materials.

- **microtome**
  an instrument for cutting thin section of specimens.

- **microvegetation**
  vegetation which can be studied under the microscope.

- **microzyme**
  a micro-organism of fermenting or decomposing liquids.

- **middle lamella**
  a layer of calcium pectate running between adjoining primary cell-walls.

- **middleman**
  a person or business firm operating between producers and consumers and performing marketing functions, a term frequently used to describe the nonfarm firms in the food industry, food processors, wholesalers, and retailers are middlemen.

■ **middleman bias**
a commonly held attitude that middlemen are not productive, and that they frequently exploit the farmers and consumers with whom they deal.

■ **midrib**
the largest vein in a leaf, running through the middle of the lamella longitudinally.

■ **mid-summer growth**
the second period of active growth shown by some trees.

■ **millibar**
a pressure unit of 1000 dynes per square cm. in recording atmospheric pressures indicated by a barometer. One bar equals 1000 mb.

■ **millimetre of mercury (mmhg)**
a unit of pressure. The pressure that will support a column of mercury one millimetre high under specified conditions.

■ **millipore filters**
a series of membrane filters commonly used to remove bacteria.

■ **millirem**
one-thousandth of a rem. EPA in 1976 allowed a maximum annual dose of 4 millirems in drinking water, a figure that was disputed by atomic power stations as being much too low.

■ **milliroentgen (MR)**
one-thousandth of a roentgen. Symbol mr.

■ **milorganite**
a by-product of the purification of sewage waters, used as fertiliser. It contains about 4-6% N and 2-4% $P2O_5$. A brownish powder.

■ **milton's reagent**
reagent used to test for proteins, made by dissolving mercury in nitric acid, diluting, then decanting the liquid from the precipitate.

■ **mine dumps**
areas covered with overburden and other waste materials from ore and coal mines, quarries, and smelters, and usually with little or no vegetative cover.

■ **mine run potash salts**
potash salts containing a high percentage of chloride from 20 to 30 per cent potash ($K_20$).

■ **mine wash**
water deposited accumulations of sandy, silty or clayey material recently eroded in mining operations. It may clog streams and channels, and damage land on which it is deposited.

- **mineral**
a natural inorganic substance whose chemical composition is either definite or varies only within prescribed limits, and which has more or less definite physical properties such as crystal form, cleavage, hardness, colour, streak, specific gravity, solubility etc.

- **mineral reserve**
the minerals of the soil rich in nutrient elements which are released gradually. Soils of dry regions usually have a greater reserve fertility. A part of these reserves disappears through leaching, proportionally to the wetness of the climate and to the porosity of the soil.

- **mineral water**
groundwater that rises to the surface through a natural opening in the earth or rock and contains a relatively high concentration of mineral ions and trace elements.

- **mineralisation**
the conversion of an element that is immobilised in some organic combination to an available form as a result of microbial decomposition.

- **mineralisation of humus**
the transformation of the organic substances into mineral or gaseous substances by subdivision of a large molecule into several small ones.

- **mineralisation of nitrogen**
the conversion of amino groups in the soil organic matter to $NH_4^+$ and $NO_3$ ions.

- **mineralogical analysis**
estimation of the kinds or amounts of minerals in soil or rock.

- **minimum temperature**
the lowest temperature recorded during a given period.

- **minimum thermometer**
a thermometer that records the minimum temperature achieved over a specified period. The most common type consists of an ordinary alcohol thermometer, fixed horizontally, with a metallic marker within the column of alcohol. The surface tension of the meniscus of the alcohol moves the marker down the tube as the temperature falls, and leaves it at the lowest point reached when the temperature rises again. Thus the position of the marker indicates the minimum temperature reached in any given period of time.

- **minimum tillage**
that amount of tillage required to create the proper soil condition for seed germination, plant establishment, and prevention of competitive growth. Herbicides are often subjected to some tillage operations. The favourable effects of minimum tillage have been put forward in recent times. They are:
a. reduces compaction,
b. reduce damage to soil aggregate,
c. reduces cost of crop production,
d. decreases oxidation of organic matter
e. reduces soil erosion
Moreover, it can greatly abate the critical problem of pollution.

- **minor elements**
also called minor plant food elements, trace elements. A term often used to refer to those fertiliser elements other than the major elements, which are essential to plant growth. Examples: Copper, zinc, boron, iron, manganese, chloride and molybdenum.

- **minor watershed**
part of a watershed which may be selected as a suitable area for detailed planning and application of soil and water conservation measures in cooperation with relatively small local groups of landowners and operators, or local agencies as drainage districts or municipalities.

- **mirror image**
a shape that is identical to another except that its structure is reversed as if viewed in a mirror, so that the two cannot be superimposed.

- **miscellaneous land type**
are used in soil classification and mapping for areas of land that have little or no natural soil or that are too nearly inaccessible for orderly examination, or where, for other reasons, it is not feasible to classify the soil. They are named primarily in terms of land form and secondarily in terms of material.

- **miscibility**
the tendency or capacity of two or more liquids to dissolve in each other, degrees are total miscibility, partial miscibility, and immiscibility.

- **miscible liquids**
two or more liquids capable of being mixed and which will remain mixed under normal conditions.

- **mispairing**
the condition of having a base in the DNA chain which does not pair with the base of the other chain of the double helix.

- **mist blower**
spray equipment in which hydraulic atomisation of the liquid at the nozzle is aided by an air blast past the source of spray.

- **mitochondria**
1. specialised subcellular structures located within body cells that contain oxidative enzymes needed by the cell to metabolise foodstuffs into energy sources.
2. organic - Being composed of, or containing matter of plant or animal origin.

- **mitosis**
the process of nuclear division in cells that produces daughter cells that are genetically identical to each other and to the parent cell.

- **mitotic index**
the number of cells simultaneously in the process of division, out of 1000 cells.

- **mitscherlich's law**
assuming that a plant produces its maximum yield if all conditions are ideal, but if an essential factor is absent, then there is a corresponding reduction in yield, and assuming an increase in yield by unit increment of the lacking factor is proportional to the decrement from the maximum, then:
Where y is the yield when x is the amount of the factor present, C is a constant.
A is the maximum yield if x is in excess, and C is a constant.

- **mix pricing**
a pricing tactic of multi-product retailers whereby each product is assigned a different gross profit, thus attempting to differentiate one retail chain store from others.

- **mixed commercial farming**
a system of farm that not only produces for own consumption but also for commercial purposes. It is a transitional stage from subsistence farming to commercialisation farming.

- **mixed cropping**
  a system of growing together more than one crop at a time on the same piece of land. Crops may be grown in a homogenous mixtures in which seeds of various crops are mixed before sowing or that different crops may be sown in separate rows according to a pattern. Mixed cropping benefits are:
  it is an insurance against complete failure of the crop,
  better utilisation of plant nutrients and soil moisture, and space.
  Following considerations govern the selection of crops for mixed cropping:
  Tall crops are mixed with short crops
  Early maturing with the late maturing ones
  Legumes with non-legumes
  Wide row crops with close row crops
  Shallow rooted and deep rooted ones.

- **modifying gene**
  a gene whose differences are revealed by their effect on the expression of another gene.

- **moist climates**
  when the moisture index (Im) is positive for a place it is said to have a moist climate.

- **moisture (volume percentage)**
  the ratio of the volume of water in a soil to the total bulk volume of the soil.

- **moisture (weight percentage)**
  the moisture content expressed as a percentage of the oven-dry weight of a soil.

- **moisture adjustment**
  adjustment of the observed precipitation in a storm by the use of the ratio of estimated maximum perceptible water, over the drainage area under study, to the actual perceptible water, calculated for a particular storm. This procedure is one step in determining the maximum probable precipitation for design purpose and may lead to a statistically derived design flood.

- **moisture density**
  the mass of water per unit volume of space occupied by soil, air and water.

- **moisture equivalent**
  1. the amount of water retained by a soil when a layer of that soil 1 cm deep is centrifuged for 40 minutes in a gravitational field equivalent to 1000 times that of gravity. This is an arbitrary but use-

ful constant as an estimate of the field capacity of soils. The soil moisture tension is about one-half of an atmosphere *i.e.* pF=2.7.
2. moisture equivalent of a soil could also be calculated from other soil moisture constants as shown below:
M.E. = (Moisture holding capacity—21) X 0.635, M.E. = Hygroscopic coefficient x 2.71, M.E.= Wilting coefficient X 1.84, M.E.=0.02 sand+0.022 salt+1.05 of clay.

- **moisture release curve**
depicts the relationship between soil moisture tension and the absolute moisture content in the range from field capacity to wilting point.

- **moisture retention curve**
a graph showing the soil moisture percentage (by weight or by volume) versus applied tension (or pressure).

- **moisture retentive soil**
a soil that retains a high percentage of water retentiveness directly depending upon clay colloid content, clay soil more retentive than sandy soil.

- **moisture tension**
the equivalent negative gauge pressure, or suction, in the soil moisture. Soil moisture tension is equal to the equivalent negative or gauge pressure to which matter must be subjected in order to be in hydraulic equilibrium, through a porous permeable wall or membrane, with the water in the soil.

- **molal quantity**
the number of moles (gram-molecular weights) present, expressed with weight in pounds, grams, or such units, numerically equal to the molecular weight, for example, pound-mole, gram-mole.

- **molal solution**
the concentration of a solution expressed in moles of solute divided by 1000 grams of solvent.

- **molecular gas**
a gas composed of single species such as oxygen, chlorine, nitrogen.

- **molecular sieve**
a substance through which molecules of a limited range of sizes can pass, enabling volatile mixtures to be separated.

- **molecular sieve hypothesis**
the theory that membranes have pores of fixed size through which particles can

pass if they are not too big.

- **mollic epipedon**
  a surface horizon that, when mixed to a depth of 17.5 cm (7 in.), contains 1% or more organic matter, with colour values darker than 5.5 dry and 3.5 moist. Structure cannot be massive and hard. Base saturation is over 50% and the epipedon is not naturally dry more than 3 months per year. It has strong structure so that the soil is soft instead of hard and massive when dry.

- **mollisols**
  soil taxonomic order consisting of semiarid grassland prairie soils of high fertility, near-neutral pH, a thick A horizon high in dark organic matter, and excellent structure for cultivation.

- **molybdenum**
  its use as an alloying element in steel increases hardenability and in low alloy steels reduces the risk of temper brittleness. When added to stainless steels it increases their resistance to corrosion. It is also used in high speed steels.
  Molybdenum has been shown to be required for nitrogen fixation by N—fixing micro-organisms. It is perhaps concerned in the enzymatic transfer of electrons required for the reduction of the Nsbonds to HN=NH bonds from molecular hydrogen.
  The poor formation of middle lemella in molybdenum-deficient plants is suggestive of a role of molybdenum in pectin metabolism. It is absorbed by plant roots as the ion $MoO_4^{2'}$.

- **molybdenum poisoning (molybdenosis)**
  the poisoning of farm animals, such as cattle, sheep and goats as a result of the presence of molybdenum in feeds or forage. Some animals resistant. Poisoning characterised by anaemia, diarrhoea, emaciation, and reduction in milk production.

- **monochromogenous soil**
  a soil derived from the direct decomposition of crystalline rocks.

- **monocot weed**
  weed producing single cotyledon seeds.

- **monocotyledon (monocot)**
  any seed plant having a single cotyledon or seed leaf. Includes corn and grass-type plants.

▪ **monoculture**
the practice of growing the same crop year after year on a particular piece of land.

▪ **monoecious**
plants with separate male and female flowers on same individual.

▪ **monopoly**
the word monopoly is made up of two syllables— *mono* and *poly*. *Mono* means single while *Poly* implies selling. Thus, if there is only one single seller of a product in the market, that situation would be referred to as monopoly.

▪ **monopsony**
it refers to the market structure with a single buyer of a commodity. Pure monopsony or buyer's monopoly is characterised by the ability of the single buyer to set the buying price. It is not very common, but it may occur, as to the case of the demand for labour in a company town. In the case of monopsony, the buying price and the quantity bought are lower than they would be in a competitive situation.

▪ **monsoon**
the term monsoon is generally applied to quasi-stationary disturbances of average zonal circulation, particularly in the tropics arising from temperature and humidity differences between continental and oceanic areas , and between northern and southern hemisphere.

▪ **montmorillonite**
is a clay mineral. Each of its layers contains two tetrahedral sheets and one octahedral sheet. The two tetrahedral sheets are held on opposite sides of the octahedral sheet by apex oxygen ions replacing hydroxyl ions as in the kaolinite structure. Some of it's aluminium may be replaced by iron or magnesium and some of the silicon may be replaced by aluminium. Variable amounts of water are absorbed between the layers. The cation substitution lead to a deficiency of positive charges in the structure with the result that the layers are negatively charged. The negative charges give montmorillonite a relatively large capacity to attract cations such as $Ca^{++}$ to the layer surfaces, but the attraction is loose enough to permit other cations to replace the ones already there in an exchange process. Montmorillo-

nite has an expanding lattice that permits both water and cations to enter between its layers. The exchangeable cations provide an attractive force between the layers but the attraction is too weak to prevent water from entering.

▪ **moonlighting**
it is holding of more than one job by an individual worker. A moonlighter may do one job during daytime and take up another in the evening hours.

▪ **moor (moorland)**
an open upland with thin peaty soil, which is wet and acidic. It typically supports a vegetation of heather, coarse grass, bracken, and moss. Such areas are usually used for rough grazing.

▪ **moorum**
a coarse grained material found as layered deposit above rock. Due to disintegration from parent rock, it contains some fine grained component also. The colour is generally red, yellow or brown.

▪ **morphine**
the principal active ingredient in opium. It is considered by some to be superior to other pain relievers.

▪ **moss**
a form that cadmium metal takes when the molten metal drops into cold water. This bubblelike form of solidified metal enhances dissolution.

▪ **mosses**
a group of small green plants with neither roots nor vessels, but with tissues, which reproduce by aerial spores formed in a sporangium.

▪ **mottled**
means marked with spots of colour.

▪ **mottled zone**
spots or blotches of soil of a certain colour or shade of colour interspersed within a soil with a different dominant colour.

▪ **mottling**
1. irregularly marked with spots of different colours.
2. colours different from those of the general soil mass.

- **mouldboard plough**
a common implement used for primary tillage operation. It has the advantage of serving purposes at a time such as cutting the furrow slice, lifting the furrow slice, pulverising the soil, turning the furrow slice.

- **mound**
a purposefully constructed circular earthwork built by prehistoric and early historic people, used primarily for the interment of the dead although some may have functioned as foundations for living structures.

- **movement of herbicide**
advancement of absorbed herbicide across the thickness of a root or shoot.

- **mowing**
the cutting of a crop close to the ground. The crop is later put through a harvester header to thresh it.

- **mucilage**
gelatinous secretions and exudates produced by plant roots and many microorganisms. It plays an important role in coating and bridging soil particles and thus maintaining and improving soil structure.

- **muck**
fairly well decomposed organic soil material, relatively high in mineral content, dark in colour, and accumulated under conditions of imperfect drainage.

- **muck soil**
consists of highly decomposed plant residues of no recognisable origin, mixed with washed-in mineral soil and/ or underlying mineral parent material.

- **mud filters**
a sugar factory term used for fine screen sepation of juice from mud on a rotating drum under reduced pressure.

- **mud rain**
rain that contains quantities of dust suspended within it. The dust originates chiefly from desert dust-storms, although sometimes it may be debris from a volcanic eruption.

- **mulch for weed control**
any material employed to cut light to the young, germinating weeds.

- **mulch planter**
a tool that destroys a narrow strip of the sod (35-50 cms) and plants the corn in this cultivated strip.

- **mulch tillage**
tillage or preparation of soil in such a way that plant residues are left on the surface.

- **mulching**
the natural and gradual decomposition of dead organic matter that has been evenly distributed in a thin layer on the ground. The object of this practice is moisture conservation, temperature control, prevention of crust formation, weed control, improvement in soil structure and reduction in runoff and erosion. Few common mulches are jowar or bajra stubbles, paddy straw or husk and saw dust.

- **multiple correlation**
a statistical technique that predicts values of one variable on the basis of two or more other variables. An equation may be developed through multiple correlation in which the dependent variable is expressed in terms of the independent variables. For example, the effect on flood season runoff, a dependent variable of such independent variables as antecedent precipitation, ground water conditions, temperature and snow depth may be determined through multiple correlation.

- **multiple cropping index**
an indicator of the number of crops harvested per year from the same parcel of land, computed as annual harvest area divided by cultivated area and multiplied by 100. The multiple cropping index is generally higher with smaller farm *size.*

- **multiple equilibrium**
exist when several different sets of prices and quantities will meet the equilibrium conditions.

- **multiple factors**
two or more factors which have similar effects on a particular character.

- **multiple gene inheritance**
determination of a quantitatively varying traits by the additive effects of more than one allele.

- **multiple plough**
a plough with two or more bottoms. Also called a gang plough.

- **multiplier**
the multiplier is the ratio of the change in equilibrium GDP (Y) divided by the original change in spending that causes the change in GDP. It was introduced into economics for the first time in 1931 by R. F.

Kahn and was later refined by John Maynard Keynes in his 'Theory of Employment'. It deals with the magnified effect that changes in investment have no income and employment.

- **mutation**
a change in the genetic information.

- **mutation pressure**
the measure of the action of mutation intending to alter the frequency of a gene in a given population.

- **mutation rate**
a proportion of individuals or cells in a given group, which show mutation for a given gene under given conditions in one generation or other stated unit of time.

- **mutual adaptation**
the adaptability of the organism and host to continue living together without mutual harm to each other.

- **mutualism**
relationship between two or more species that benefits all parties to the relationship, a benevolent association between two microorganisms that benefits both, a symbiotic relationship in which both members benefit. Neither can survive without the other.

- **mycelium**
a complex network of thread-like tubular filaments (hyphae) possessed by many fungi that function in nutrient absorption and transfer.

- **mycorrhiza**
a mutually beneficial (symbiotic) association between a plant root and a fungus, that enhances the ability of the root to absorb water and nutrients.

- **myxo**
a prefix meaning mucoid or mucus.

- **n serve**
a trade name for chemical 2-chloro-6 (trichloro-methyl) pyridine which is used as nitrification inhibitor. The recommended way of application is by mixing it with ammonical fertiliser. The rates used are about 1-2% N-serve by weight of fertiliser nitrogen.

- **n2**
symbol for slightly leached brown forest soil.

- **narrow base ridge terrace**
type of contour terrace for soil conservation. Commonly called bund, and consists of a comparatively narrow-based

embankment constructed at intervals across slope of land either on a level (along the contour) or slightly graded longitudinally, if conditions require removal of excess water. Vertical spacing between bunds varies from 3 to 6 ft. depending upon slope, type of soil, rainfall. Constructed on agricultural land up to a slope of about 10 per cent.

- **narrow base terrace**
a terrace similar to a broad-base terrace in all respects excepting the width of ridge and channel, the base width of a narrow terrace is usually from 4 ft to 8ft.

- **narrow leaf species**
those plants classified as monocotyledonae.

- **native pastures**
a term referring to native vegetation in an untilled prairie or plains area. The term 'unimproved' is used instead of 'native' for the pastures which include mainly species not native.

- **natric horizon**
a subsurface diagnostic soil horizon. This horizon meets the requirements of an argillic horizon but also has prismatic or columnar structure and over 15% of the CEC is saturated with $Na^+$, or has more exchangeable plus $Na^+$ than $Ca^{++}$ plus exchange acidity at pH 8.2.

- **natural**
composed of naturally occurring materials, or created, or appearing to have been created, by natural processes.

- **natural classification**
a classification that groups organisms or objects together on the basis of the sum total of all their characteristics, and tries to indicate evolutionary relationships.

- **natural combustion**
combustion of organic substances on the surface of the soil or in the soil, with intervention of microorganisms. It is favoured by light and heat and slowed down by drought or an excess of water.

- **natural control**
nature's method of controlling plant and animal populations through the action of heat, cold, moisture, dryness parasites, predators, disease, etc.

- **natural economy**
wearing away of the earth's surface by water, ice, or other natural agents. Under natural environmental conditions of

climate, vegetation, etc., undisturbed by man.

- **natural ecosystem**
  an eco-system that ultimately produces the climax community, which develops in the absence of any major disturbance and consists of a complex, self regulating food web of several tropic levels.

- **natural erosion (geological erosion)**
  erosion of the natural landscape undisturbed by man or domestic animals.

- **natural grassland**
  a category of vegetation region, occurring in areas which experience a season of prolonged drought, but having some rains during the period of growth.

- **natural group**
  a group of organisms having a common ancestor.

- **natural immunity**
  immunity to disease that occurs as part of an individual's natural biologic makeup.

- **natural subirrigation**
  the delivery of water to plant roots by the capillary fringe.

- **natural subsurface irrigation**
  application of irrigation water below surface of soil. Possible where an impervious layer exists below the root-zone, where water is led to a series of ditches dug up to the impervious layer, which then moves laterally and wets the root zones, idea being to raise the water table so that plant roots can get water by capillary movement.

- **natural ventilation**
  process of supplying or removing air using the natural forces of wind and temperature difference to or from any space, such as interstitial space between products, used in connection with conditioning of harvested produce.

- **natural water table**
  a water table in its natural condition and position not disturbed by artificial additions or extractions of water.

- **naturalisation**
  the introduction of animals or plants to places where they flourish but are not indigenous.

- **naturalised plants**
  plants introduced from other environment, which have become established in, and more or less adapted to, a given region by being grown for many generations, reproduc-

ing freely by seed and maintaining its position in competition with indigenous plants.

- **nature**
biodiversity, landforms, geology, aquatic features and atmosphere together with the dynamic interactions between these components.

- **necrobiosis**
(physiology) the normal degeneration and death of living cells (as in various epithelial cells).

- **necrosis**
the death of an individual cell or groups of cells in living tissue. Sometimes seen in carcinomas.

- **nectar**
nectar is the sugary liquid produced by many flowers. Butterflies, fruit bats, ants and gliders love to drink nectar.

- **negative electron**
change in the state of a system that counteracts the measured effect of the initial alteration.

- **negative feedback**
tendency of a system of counteract externally imposed change and return to a stable state.

- **negligible residue**
a tolerance which is set for a food or feed crop which contains a very small amount of pesticide at harvest as a result of indirect contact with a chemical.

- **nematocide**
chemical pesticide that kills nematodes.

- **nematodes**
are small worms that represent a higher form of animal life than protozoa. Most of them live on decaying organic matter but some infect plant roots and live as parasites. Some do serious damage to the roots of fruit trees.

- **neo-darwinian**
theory of evolution through natural selection of Darwin modified and expanded to modern genetic studies arising from the work of Mendel. Such studies have answered many questions which Darwin's theory raised but could not adequately explain because of the lack of knowledge at the time it was formulated.

- **neo-lamarckism**
  modification of Lamarck's doctrine of evolution, that inherited acquired characters formed inception of specific differences.
- **neo-malthusianism**
  limitation of family size by artificial birth-control.
- **neosol**
  a soil of recent formation.
- **net weight**
  weight of the goods alone without any immediate wrappings, e.g., the weight of the contents of a tin can without including the can's weight.
- **neutral equilibrium**
  the state in which a disturbance may cause a body to be displaced from its rest position, but left alone the body is neither accelerated toward nor away from the original position.
- **neutralise**
  make chemically neutral.
- **neutrobasophile**
  a plant which grows well on neutral or basic soils having pH ^ 7.
- **neutron moisture meter**
  a unique method of determining soil moisture in the field which involves neutron scattering. The neutron moisture meter is based upon the principle that hydrogen is relatively unique in its ability to drastically reduce the speed of fast-moving neutron and to scatter them. The probe, containing a source of fast neutrons and a slow neutron detector, is lowered into the soil through an access tube. Neutrons are emitted by the source (for example, americium-beryl-lium or radium) at a very high speed. When these neutrons collide with a small atom such as hydrogen contained in soil water, their direction of movement is changed and they lose part of their energy. These slowed neutrons are measured by a Ninhydrin a reagent used to test for the presence of proteins and amino acids. An aqueous solution turns blue in the presence of alpha amino acids in solution. When dissolved in an organic solvent it is used as a developer to colour amino acids on chromatograms. If a chromatogram treated with ninhydrin it is heated strongly, the amino acids appear as purple spots that can be identified by measuring the Rf value.

- **NI**

symbol for leached brown forest soil.

- **nipping**

an act of removing of ear heads from plant stalk.

- **nitrate reductase**

a enzyme that catalyses the reduction of nitrate to nitrite. The enzyme system includes a reduced pyridine nucleotide (nadph or nadh) as an electron donor, flavin adenine dinucleotide(FAD) as a prosthetic group, and molybdenum as an activator. Electrons are passed from reduced pyridine nucleotide to molybdenum producing reduced molybdenum which in turn passes electrons to nitrate reducing it to nitrite.

- **nitration**

1. the conversion of nitrites to nitrates by soil bacteria, followed by the uptake of the nitrates by higher plants and their incorporation into amino acids, etc.

2. a general term for the process in which a substance is nitrated, *i.e.* has an $NO_2$ group added to it.

- **nitric phosphate**

is a granulated product containing 20% N and 8.7% P. Crushed phosphate rock is reacted with nitric acid and phosphoric acid in agitated extractors. The resulting slurry flows to a preneutraliser where 75 to 85 % of the required ammonia is added. The slurry is fed onto a recycle bed in an ammoniator-granulator where the remaining ammonia is added and granulation takes place. The product is dried, cooled, and screened to give the desired product specification.

- **nitrification coefficient**

the measurement of the production of nitric acid produced by humus during a known time in an incubator. It is compared with the total nitrogen.

- **nitrification inhibitors**

are nonphytotoxic chemicals which selectively inhibit nitrification or, to a lesser extent, those which retard denitrification or urease activity. Many chemicals inhibit nitrification to varying degrees, but few have all of the desired characteristics of specificity, mobility, persistence, and economy.

Chemicals which meet these specifications and selectively inhibit *Nitrosomonas* activity are safe to use. When added to fertilisers, they will prevent rapid nitrification and remain effective throughout the fertiliser soil reaction zone. These chemicals are thus effective in reducing nitrogen losses through leaching and denitrification and improve the recovery of applied nitrogen.

- **nitrogen deficiency symptoms**
under conditions of nitrogen deficiency the leaves are small, the stems are thin and upright, and lateral shoots are few, hence growth has a sparse appearance. The leaves usually have a pale, yellowish green colour in the early stages of growth and may develop, yellow red or purple colours as they grow older. Symptoms are most pronounced on older leaves. Under conditions of severe deficiency, the leaves ultimately become yellow and are shed.

- **nitrogen excesses**
1. the presence of nitrogen as nitrates or ammonium salts in the soil in excessive amounts as to cause injury to plants.
2. condition brought about by plants-feeding and absorbing nitrogen in excess of their requirements. Produce leathery, sometimes crinkled dark green leaves and succulent growth, delay maturation of plants, impair quality of crops like barley, potato, tobacco and fruits, increase susceptibility to disease and cause lodging of cereal crops by inducing an undue lengthening of the stem internodes.

- **nitrogen fertilisers (natural organic)**
acidulated fish, tankage, acidulated fish scrap. The rendered product derived from fish and treated with sulphuric acid.

- **nitrogen fertilisers (synthetic organic)**
a commercial product consisting principally of calcium Cyanamid ($CaCN_2$) and carbon. It contains not less than 20 per cent nitrogen and 54 per cent calcium oxide. Urea—the commercial synthetic acid amide of carbonic acid. It contains not less than 42 per cent nitrogen.

- **nitrogen fixation**
1. the conversion of atmospheric nitrogen to nitrogen compounds that can be used

by plants. When brought about by bacteria in the root nodules of leguminous plants, it is spoken of as symbiotic, if by free-living micro-organisms acting independently, it is referred to as non-symbiotic fixation.
2. the chemical and physical processes used by fertiliser plants to convert nitrogen gas to ammonia.

- **nitrogen fixing plant**
a plant which can assimilate and fix the free nitrogen of atmosphere by the aid of bacteria living symbiotically in the root nodules.

- **noculum potential**
energy of growth of the pathogen available for infection of the host at the surface of the host to be infected.

- **node**
the joint of a culm where a leaf is attached.

- **nodulation**
the production of nodules on roots of legumes by specific *Rhizobium* bacteria.

- **nodule**
a spherical, oval or similarly rounded concretion not exceeding 256 mm in diameter. The term nodule and concretion are often used synonymously, nodule being preferred for more rounded types.

- **nodule root**
a tubercle-like formation on certain leguminous plant roots attributable to the invasion of symbiotic nitrogen-fixing bacteria. The bacteria furnish the plant with fixed nitrogen compounds (nitrogen taken from the air, and in return receive nutrient plant juices like carbohydrates). The genus Rhizobium and some species of *Azotobacter* and *Clostridium* fix free nitrogen.

- **nodules (glaebules)**
swellings on root of most Fabaceae, produced by bacteria (e.g. Rhizobium spp.) associated with the roots of these plants. These bacteria are able to fix atmospheric nitrogen and to accumulate it in the nodules. Hydroxides are probably formed by accretion (plasma concentration).

- **nomogram (nomograph)**
a graphic representation of

numerical relations, without calculation, from the values of one or two other variables which are known.

- **numerical hybrid**
  a hybrid whose parental gametes differed in respect of the number of chromosomes.

- **numerical mutation**
  a change in the number of chromosomes, either balanced to give polyploidy, or unbalanced to give aneuploidy.

- **nutrient requirement**
  the minimum concentration of nutrients required in a tissue to allow growth.

- **nutrient spray**
  a liquid fertiliser (such as urea) or plant food sprayed directly on the foliage of a plant from which it is absorbed.

- **nutrient toxicity**
  poisonous effect produced in plants by application in excess of food element requirements, more so in the case of micronutrient, chlorosis and defoliation in citrus and dry rot in beets produced by excess of boron, golden yellow shoots in tomatoes by excess of molybdenum, stem streak necrosis in potato by excess of manganese.

- **nutrients and food**
  the term nutrient refers, generally to raw inorganic materials from which complex organic compounds called foods are synthesised. Foods are store houses of energy. Examples of nutrients are nitrogen, phosphorus etc. and those of foods are carbohydrates.

- **nutrition**
  nutrition is the process of nourishing, by which organisms obtain energy (in the form of food) for growth, maintenance, and repair.

- **nutritional labeling**
  labels that provide consumers with information about products' nutritional value.

- **nyirok soils**
  red soils of Moravia and Hungary, with affinities to a terrarosa, but may have resulted from Tertiary weathering.

- **o horizon**
  organic soil horizon formed from organic litter derived from plant and animal and deposited on the mineral surface. These horizons are measured upward from the top of the underlying mineral material *i.e.* 15 to 0 cm.

- **objectionable weed**
  it refers to those weed species, whose seed is difficult to sepa-

rate once mixed with crop seed or which is poisonous or injurious or has a smoothering effect on the main crop. It is difficult to eradicate once established because of its high multiplication ratio. It may also serve alternate host for crop disease and pests.

- **objectivity coefficient**
  the correlation coefficient between two independent ratings or scorings of the same set of products by the same judge.

- **obligate**
  restricted to a particular condition of life, as an obligate seeder, a plant that can only reproduce by seed. Thus, *e.g.* an obligate aerobe grows only in the presence of free oxygen, an obligate parasite is one which can not exist independently of a host.

- **oil bearing crops**
  crops grown primarily for their oil content, such as soybean, groundnut, cotton seed, linseed, flax, mustard, sunflower, safflower sesame, rape, etc.

- **oil bordeaux mixture**
  is prepared by the addition of edible cotton seed oil to bordeaux mixture. It improves the retention of copper on foliage.

- **oil cakes**
  are by-products of oil mills. These are rich quick acting nitrogenous manures.

- **oil emulsion**
  contains lube oil (2 gallon), water (1 gallon), Potash fish oil soap (2 pounds)—was used in the beginning of 20th century for the control of pests.

- **oil waste land**
  includes areas where liquid oily wastes have accumulated. This miscellaneous land type includes slush pits and adjacent uplands and bottoms affected by the liquid wastes, principally salt water and oil. The land is unsuited to agricultural purposes, although some of it can be reclaimed.

- **oilseed**
  any of several seeds that yield oil.

- **old fallows**
  cultivable lands not cultivated for a number of years.

- **oligocleraent**
  an element which, in very small quantities, exerts a favourable action on plants, *e.g.* Mn, B, Zn, Cu, Mo, Cl etc.

- **oligopsony**
  a market structure in which there are relatively few buyers. An oligopsonistic situation may lead to express or tacit collusion among the buyers to depress their buying prices, generally at the expense of the sellers who supply them. An oligopsonistic market may face either an atomistic, competitive selling market or one with a significant degree of seller concentration.

- **oligotrophophyte**
  a plant growing in a soil poor in soluble mineral salts.

- **ombrophile**
  a plant which thrives in a place where rain is abundant.

- **ombrophobe**
  a plant which cannot survive in continuous rain.

- **ombrophyte**
  a plant inhabiting rainy places.

- **omnivorous**
  said of a parasitic fungus which attacks several or many species of host-plant.

- **on going evaluation**
  an evaluation undertaken during the implementation phase of a project.

- **on the dry side**
  keeping soil or planting medium barely moist as compared to rather wet. A recommended procedure to keep losses from seed rot, damping-off and seeding blight at a minimum.

- **one tailed test**
  a statistical test of significance involving only a single side tail of the sampling distribution.

- **one third atmosphere moisture point**
  the percentage of moisture retained by a soil sample when placed on a porous plate and subjected to a pressure of 1/3rd atmosphere (pF= 2.54). It is supposed to be a good estimate of field capacity and moisture equivalent. Sometimes called the upper limit of readily available water.

- **one way**
  a tillage implement having a gang of disks that through the soil in one direction.

- **one way plough**
  type of mouldboard plough which throws the furrow slice to one side of the direction of motion.

- **open economy**
  an economy in which there are no restrictions on trade with

areas outside that economy, as opposed to a closed economy in which there is no foreign trade.

- **over irrigation**
the application of more water than is necessary for the needs of vegetation, resulting in loss of water through seepage and leaching with the resultant loss of humus, nitrogen and other mineral elements of the soil.

- **overburden potential**
it exists wherever soil is free to move and its weight becomes involved as a force acting upon water. It behaves similarly to a pressure potential that is caused by the weight of water above and sometimes included as an implicit part of a pressure potential.

- **overcalcification**
enrichment of the parent rock in lime by concentration formation.

- **overconsolidated soil deposit**
a soil deposit that has been subjected to an effective pressure greater than the present effective over-burden pressure.

- **overland flow (surface flow)**
the movement of water in films, sheets, or rills across the surface of the ground. It occurs during heavy rainfall when the soil cannot absorb any more water as all pore spaces are full (*i.e.* the infiltration capacity is exceeded). Overland flow is most noticeable in arid and semi-arid areas where there is little vegetation. As rainfall continues in such areas the sheet flow may develop into rills and then into gullies causing considerable erosion.

- **overland runoff**
water flowing over the land surface before it reaches a definite stream channel or body of water.

- **overlapping system of cropping**
a system of multiple cropping in which the succeeding crop is sown in the standing crop, *e.g.* Paddy-Lathyrus, Paddy-Lucerne, Paddy-Berseem, Cotton-Berseem.

- **oversaving**
the theory that oversaving may result in under consumption. J. A. Hobson believed that oversaving was the cause of the downswing of the business cycle. The main cause of oversaving, he believed, was inequality of incomes.

- **overstocked**
condition of overcrowding in tree crops, leading to retarded growth.

- **oxamide**
it is a slow release material and source of N for plants. Oxamide, the diamide of oxalic acid, is a white, non-hygroscopic compound which contains 31.8% N. It is not yet commercially produced as fertiliser. The process that currently appears most feasible requires first the production of cyanogen from hydrogen cyanide by either of two routes. The cyanogen is then hydrolysed in the presence of concentrated acid to oxamide.

- **oxic horizon**
an altered subsurface horizon consisting of a mixture of hydrated oxides of iron or aluminium and 1:1 clays and/or quartz.

- **oxidation number**
numerical charge on the ions of an element.

- **oxidation reduction indicator**
a compound whose colour in the oxidised state differs from that in the reduced state.

- **oxidation reduction reaction**
an oxidising chemical change, where an element's positive valence is increased (electron loss), accompanied by a simultaneous reduction of an associated element (electron gain).

- **oxidation state**
the number of electrons to be added (or subtracted) from an atom in a combined state to convert it to elemental form. Also known as oxidation number.

- **oxide cutan**
a kind of cutan consisting of a thin layer of a metal oxide, usually of iron (sesquan) or manganese (mangan). The colour, usually different from that of the ped interior, is red where iron oxide is prominent and black where manganese is an important constituent.

- **oxy**
prefix indicating the oxygen radical (—O—) in a chemical compound.

- **oxyhydrophyte**
vegetation of temporarily submerged acid soils.

- **oxyphile**
a plant which grows well in acid soils. Oxyphyte. The

plant formation of acid soils.

- **p**
a sub-horizon symbol for ploughing or other disturbance. The symbol p is used as a suffix with A to indicate disturbance by cultivation or pasturing. Even though a soil has been truncated and the plough layer is clearly in what was once B horizon, the designation Ap is used.

- **package plant**
a prefabricated or prebuilt waste water treatment plant.

- **pakihi soil**
a soil developed under podsolic conditions and impeded drainage. Found in parts of New Zealand.

- **palatability**
the state of being agreeable to the taste.

- **palatable**
good to the taste, not necessarily healthful. Some poisonous plants that livestock usually avoid are made more palatable by treatment with certain chemicals and may cause injury to the animal.

- **paleosol**
soil which formed formerly, under conditions .generally different from those of the present day.

- **palmate**
when lobes of a leaf spread from the same point, like the fingers of a hand.

- **paludisation**
process regarded by some workers as geogenic rather than pedogenic, including the accumulation of deep (>30 cm) deposits of organic matter as in mucks and peats.

- **pan formation**
a relatively thin and hard layer or stratum of more or less impervious material occurring in the subsoil, and frequently extending over a considerable area, usually occasioned by the accumulation of salts of iron, calcium, magnesium and other bases where there is no free drainage through the soil.

- **panicle**
a group of flowers borne at

unequal distances from .the central stem, as in orchard grass.

- **papules (glaebules)**
are glaebules composed dominantly of clay minerals with continuous and or lameller fabric. They occur as inherited lithorelicts, pedorelicts or sedimentary relicts or any disrupted orthic pedological features (*e.g.* disrupted cutans). They have sharp external boundaries. Most commonly, they are prolate to equant and somewhat rounded.

- **parabolic**
having a broad base, and gradually narrowing by curved side to a blunt point.

- **photometry**
the branch of Physics which deals with the measurement of the amount of light energy emitted or absorbed by a body.

- **photomicrograph**
a photograph of the grain structure of a metal as observed when optically magnified more than 10 diameters. The term micrograph may be used.

- **photomorphosis**
a change in the structure of a plant after exposure to strong light.

- **photophosphorylation**
production of high energy phosphate (ATP) from inorganic phosphorus and ADP through the utilisation of light energy in photosynthesis.

- **photorespiration (pr)**
a light stimulated oxidation of photo-synthesis intermediates to CC>2 and is attributed to the oxidation of glycolic acid which is biosynthesised rapidly by chloroplasts only during photosynthesis. Peroxisomes may contain enzymes of glycolate metabolism. In general, plants with pr are temperate. Tropical grasses which have $C_4$ pathway of photosynthesis lack pr and are characterised by low $CO_2$ compensation point.

- **photosynthesis gross**
the total amount of food synthesised by a green plant from coa and water in a given amount of time.

- **photosynthesis net**
the net amount of food synthesised by a green plant in given amount of time after subtracting the amount used in respiration during the same time.

- **photosynthetic bacteria**
  a group of bacteria able to photosynthesise through possession of a green pigment, bacteriochlorophyll, slightly different to the chlorophyll of plants. They do not use water as a hydrogen source, as do plants, and thus do not produce oxygen as a product of photosynthesis, but rather some oxidised by-product. Photosynthetic bacteria include the green sulphur bacteria, purple sulphur bacteria, etc.

- **photosynthetic capacity**
  the efficiency of a plant cell or a chloroplast in carrying out photosynthesis.

- **photosynthetic efficiency**
  the ratio of light-energy absorbed by the chloroplasts of a tissue in unit time, to the amount of energy fixed per unit time.

- **photosynthetic number**
  the ratio between the number of grams of carbon dioxide absorbed per hour by a unit of a leaf to the number of grams of chlorophyll which that unit contains.

- **physical properties (of soils)**
  those characteristics, processes, or reactions of a soil which are caused by physical forces and which can be described by, or expressed in physical terms of equations. Sometimes confused with and difficult to separate from chemical properties, hence, the terms 'physical-chemical' or 'physicochemical'. Examples of physical properties are bulk density, water-holding capacity, hydraulic conductivity, porosity, pore-size distribution, etc.

- **physical treatment**
  any water treatment process involving only physical changes or applied forces to obtain solid-liquid separation.

- **physical weathering**
  the breakdown of rock and mineral soil into smaller fragments by physical forces such as frost action.

- **physiogenic disease**
  a disease produced by some unfavourable physical or environmental factor, *e.g.*, excess or deficiency of light, water, chemical, plant nutrients, temperature, physical injury, etc.

- **physiography**
the study of the surface forms of a region.

- **physiological balance**
the balance between the ions in a culture solution, so that none of them can have an adverse effect on the growth of a plant.

- **physiological drought**
the state of a soil which retains water strongly and which does not release it to plants because of high concentration of soluble salts, or the retention of moisture by the organic matter, or because of low temperature, or water logging, etc.

- **phyto**
prefix denoting 'plant', *e.g.*, a phytocide is a substance that kills plants.

- **phytoalexins**
in some cases, plants are able to respond defensively to the invasion by a pathogen through the production of new fungi-toxic substances that are formed only when the host cells come into contact with the parasite. Such compounds are called as phytoalexins.

- **phytobland oil (crop oil)**
non-phytotoxic oil used to aid herbicide penetration in plant.

- **phytochrome**
a reversible pigment system of protein nature found in the cytoplasm of green plants, it is associated with the absorption of light that effects growth, development, and differentiation of a plant, independent of photosynthesis, *e.g.*, in the photoperiodic response.

- **phytogenic soil**
a kind of soil whose main factor in the formation is the ancient or present-day vegetation.

- **phytophagous**
(of animals) feeding on plants.

- **phytophasic irrigations**
refers to irrigating the crop at certain physiologically critical stages of plants.

- **phytotomy**
poisonous to plants.

- **piche evaporimeter**
it consists of a graduated tube 1.5 cm in diameter and 30 cm long with one end open. A dried paper held in place by a metallic device covers the open end. The tube is filled with water and turned upside down. The water slowly wets the paper and evaporates. Water loss is read on the tube, as a measurement of evaporation.

- **picking**
  a manual method of harvesting the field crops in which useful parts are picked as in cotton, pod fruits, etc.

- **pickling**
  seed treatment with suitable fungicides as a disease control.

- **pickup**
  an attachment for a combine or other implement to gather cut crops from a windrow and convey them to the machine.

- **picota**
  a water lifting device consisting of a wooden beam mounted on a vertical pillar in a see-saw fashion. A bucket with a long rope is tied to one end of the beam and a heavy stone is fixed to the other end. This lift makes use of human labour and is used in shallow wells of depth around 15 feet.

- **pictograph**
  a representation of an object, action or idea by a picture or symbol suggestive or imitative thereof.

- **piezometric level**
  the elevation to which the water level rises in a well that taps an artesian aquifer.

- **piezometrie surface**
  the theoretical level to which water should rise under its own pressure if tapped by a well or spring, the water level of an artesian well is one point on such a surface. An imaginary surface that everywhere coincides with the static level of the water in the aquifer. The surface to which the water from a given aquifer will rise under its full head.

- **prechill**
  placing the seeds on a moist substratum and holding them at a low temperature for a definite period before they are transferred to a higher germination temperature. The temperature employed is usually 10°C.

- **precipitable water**
  total water vapour contained in an atmospheric column of unit cross-sectional area expressed in terms of depth of water of the same cross-sectional area.

- **precipitated bone**
  the by product from m glue industry obtained by treating hydrochloric acid extract of the bone with lime.

- **precipitated phosphate**
  a product consisting mainly of dicalcium phosphate obtained by neutralising, with calcium hydroxide, the acid

solution of either phosphate rock or processed bone.

- **precipitation (p)**
precipitation is defined as water in liquid or solid forms falling to the earth. The common precipitation forms are rain, drizzle, snow, sleet and hail.

- **precipitation deficiency**
the amount by which the precipitation during a given period falls short of equalising the normal precipitation for the period.

- **precipitation/evaporation ratio**
moisture that falls from the atmosphere as rain, drizzle, hail, sleet or snow. If less, it is a dry month and adequate moisture will not be available for crops.

- **predator**
an animal that consumes other animals to satisfy its nutritive requirements.

- **predictability**
the quality of being predictable.

- **prediction**
the act or process of forecasting, with greater or less probability, the outcome of an event or series of events, by inference from a scientific, especially a statistical, analysis of known events. In view of the wide variation in the intensity of known causal factors and the presence of unknown causal factors, prediction is usually made on the assumption that certain major and unknown factors continue to operate in a given manner.

- **predispose to chemical injury**
to increase the chances of damage or harm by careless handling and incorrect use of pesticide chemicals. One must read the label and follow directions carefully to use the pesticide chemicals safely.

- **preemergence spraying**
the spraying of the soil (usually with herbicides) after the crop has been planted and before the seedlings have emerged.

- **preemergence weedicide**
weedicide applied to soil before crop seedlings appear.

- **preirrigation**
the application of irrigation water prior to the first irrigation after emergence of plants, pre-sowing irrigation.

- **premature seeding**
bolting, development of seedstalk before the usual time or

before maturity of other parts of a plant. Frequently results in losses, especially in celery, sugar beets, spinach, tobacco, etc.

- **premixing**
the preliminary blending of ingredients used at low level with or without diluents (carriers).

- **preparatory tillage**
various tillage operations carried out on the farm before the crops are sown, planted or transplanted to make a fine seedbed. Includes number of operations for preparing the land for sowing. These operations are: ploughing, clod crushing, levelling, discing, harrowing, manuring and compacting, etc. Also helps in burying the stubbles of the harvested crop in the soil to encourage their decomposition and to add to the organic matter content.

- **preplant**
application of a pesticide prior to planting a crop.

- **preplanting**
any time before crop is plated.

- **preplanting herbicide**
a herbicide that is applied before the crop is planted for the control of weeds.

- **preplanting treatment**
any treatment given to the seed, soil, planting material before the crop is planted.

- **pressmud**
a by-product of sugar industry rich in phosphorus and lime content. It is generally used in reclaiming sodic soil.

- **pressure membrane**
a membrane, permeable to water and only very slightly permeable to gas when wet, through which water can escape from a soil sample in response to pressure gradients.

- **pressure plate and membrane apparatus**
the apparatus consists of ceramic porous pressure plates or membranes of high air entry values contained in airtight metallic chambers strong enough to withstand high pressure (15 bar or more). The apparatus enables the development of soil water characteristic curves in the higher range of metric potential (> 1 bar) which is not possible in suction plates.

- **pressure system**
method of air movement for drying in which the air is forced through the product with the air duct or ducts be-

ing at a pressure above atmospheric pressure. Also called pushing or forcing system of air movement, used in connection with space or place for keeping harvested goods or products.

- **previous crop**
  the crop grown in the season immediately preceding the one in which the seed crop is grown.

- **price**
  the amount of money charged for a product or service, or the sum of the values that consumers exchange for the benefits of having or using the product or service.

- **price ceiling**
  a price below equilibrium legislated by the government which results in a shortage.

- **price consumption curve**
  a curve on a diagram showing how the consumption of a commodity changes when its price changes but consumer's income remaining unchanged.

- **price discovery**
  the processes by which buyers and sellers attempt to arrive at the equilibrium price consistent with supply and demand conditions.

- **price discrimination**
  a marketing technique which allocates the total supply of a commodity to alternative elasticity markets in a way that will maximise total seller returns. This means the charging of different prices from different groups of customers for a similar commodity or service.

- **price effect**
  when there is a change in the price of a commodity on its purchase is known as the price effect.

- **price elasticity**
  the relationship of changes in quantity supplied, or demanded, to price changes.

- **price floor**
  a legally imposed price which is above the equilibrium price level.

- **price taker**
  any firm that cannot influence the general level of commodity prices by altering the quantity produced, price-takers are often called quantity adjusters because their chief decision is to adjust output to a given price perfectly com-

petitive firms are price-takers.

- **pricing efficiency**
refers to the capability of prices to allocate resources efficiently and in accordance with consumer preferences.

- **primary host**
a host in which a parasite lives for much of its life cycle and in which it becomes sexually mature.

- **primary humus**
very poorly decomposed humus, the structure of which is that of the plant debris.

- **primary increase**
the increase in the size of a stem or root, brought about by the addition of cells from the cambium.

- **primary production**
the production of a crop itself in contrast to a product derived from it.

- **primary protein structure**
the number of polypeptide chains in a protein, the sequence of amino acids within them, and the location of inter- and intrachain disulphide bridges.

- **primary root**
the root developing directly from the radical.

- **primary sample**
a primary sample is a small portion taken from one point in the lot.

- **primary soil**
a soil developed through the disintegration and decomposition of rocks in place.

- **primary succession**
a succession starting from bare soil.

- **primary tillage**
the initial tillage operation to reduce soil strength, cover plant material and rearrange soil aggregates.

- **primary treatment**
wastewater treatment processes usually consisting of clarification and other processes to accomplish liquid-solid separation. It is the first stage in waste water treatment in which substantially all floating or settle solids are mechanically removed by screening and sedimentation.

- **primary valence bonds**
strong chemical bonds of the covalent or ionic type.

- **primary wall**
the first product in cell wall synthesis. The primary wall remains thin and elastic when the cell is enlarging.

- **primitive**
not specialised, the early or beginning type of stage.

- **primitive soil structure**
a structure which appears unrecognised.

- **primordium**
an organ in its earliest stage of development.

- **principle of accumulating differentia**
is a multiple category classification system (soil), differentiating characteristics accumulate or pyramid from the higher levels of generalisation to the lower levels. As a result, classes at the lower levels are defined and differentiated not only by the differentiating characteristic(s) used at a given categorical level but also by those which have been used as differentia at the higher levels. In the lowest category, a large number of differentiating characteristics have been accumulated such that the classes are quite narrowly and completely defined.

- **principle of comparative advantage**
an economic principle which holds that economic gains are to be had, in the form of reduced costs of production and/or increased standards of living, if under free trade conditions, each nation will specialise in and export those products it can produce relatively most efficiently, by virtue of its resource endowment, and import commodities for which it has a comparative disadvantage.

- **principle of output**
in addition to the reduction in the demand for a factor resulting from the substitution effect, there is generally a further reduction in its demand because of the curtailment in the firm's output resulting from the new equilibrium between marginal revenue and the higher level of marginal cost.

- **pyro**
prefix denoting fire, strong heat. In chemical nomenclature, it denotes a substance obtained by heating, *e.g.*, pyroboric acid, obtained by heating boric acid. It is also

used to indicate that the water content of an acid or salt is intermediate between that of the ortho- and met a compounds of the same name.

- **pyrology**
  the scientific art of protecting forests from fire.
- **Q**
  symbol for the flow, or flux of water.
- **qualitative analysis**
  the use of systematic procedures for deriving meaning from qualitative information.
- **quality of irrigation water**
  all irrigation waters contain dissolved salt. The salt content varies from 0.1 to 5.0 tons or more per acre-foot. A water quality test may alert the farmer to two hazards—the presence of salinity or the possibility of the formation of sodium soil. It also can reveal the presence of excessive amounts of bicarbonate, or boron, or both. Bicarbonate is harmful if the amount contained in irrigation water is high in relation to calcium and magnesium. Although boron in small quantities is essential to plant growth, it is toxic to many plants at concentrations only slightly above those needed for normal growth.
- **quality seed**
  good quality seeds are genetically true to-species, or variety, have capacity for high germination, are free from diseases and insects, are free from mixture with other crop seeds, weed seeds and inert and extraneous material. The germination and purity of the seed can be determined by conducting a seed test on a small representative sample drawn from the seed lot in question.
- **quasi rent**
  this term was first used by Alfred Marshall to describe rent acing to any factor of production other than land.
- **quiscent zone**
  the terminal part in the apical meristem of the root apex, where mitotic divisions are absent. May become meristematic if other part of meristem is injured.
- **R**
  underlying consolidated bedrock, such as granite, sandstone, or limestone. If presumed to be like the parent rock from which the adjacent overlying layer or horizon was formed, the symbol R is used alone. If presumed to be

unlike the overlying material, the R is preceded by a Roman numeral denoting lithologic discontinuity.

- **rabat**
  a local term in North India for Persian wheel which is a device to lift water from the open wells using a chain of buckets.

- **rabi**
  1. a crop season which generally begins in October-November and continues up to March-April.
  2. the post-monsoon season in which crops are grown mostly on moisture stored in the soil with or without significant winter rainfall.

- **radioactive isotope**
  an isotope with an unstable nucleus that stabilises itself by emitting ionising radiation.

- **radioactive pollutants**
  either artificial or natural radio-nuclides that produce, or that potentially produce, radiation that is injurious to part or all of the ecosystem.

- **radioactive series**
  a succession of radioactive nuclides, one decaying to the next until a stable nucleus is formed.

- **radioactive soil**
  soil which possessed a minute degree of radioactivity. Radioactive phosphorus and other nutrient elements are sometimes added to the soil to study their functions in plant growth.

- **radiobiology**
  the study of the effects of radioactivity on living cells and organisms.

- **radiocarbon dating**
  a dating method used to determine the age of samples containing carbon. The method measures the disintegration of the 14C atom. 14C is produced in the atmosphere by cosmic ray bombardment, and has a half-life of 5,570 years, making it useful for dating samples in the range of 0-40,000 years.

- **radiochemistry**
  those phases of chemistry involving the application of radioisotopes in the study of chemical problems as well as the study of the chemical behaviour of small quantities of radioactive substances.

- **radioecology**
  the study of radiation as affecting the relationship between living organisms and

environment, and of the ecological effects and destination of radioisotopes.

- **rain**
Precipitation in the form of liquid drops the diameter of which must be .5 millimetres or greater. Implies a steady precipitation that might last for several hours.

- **rain forest**
an equatorial forest where the rainfall is in excess of 80 to 100 inches.

- **rain gauge**
is a calibrated container that measures the amount of rainfall occurring during a specific period of time.

- **rain maker**
a rain-simulator used for applying water in drops upon a field surface.

- **rainbelt**
a region of heavy or above average rainfall.

- **rainfall**
a fall of rain—the amount of water that falls as rain expressed in units of length (depth) *e.g.*, cm.

- **rainfall distribution coefficient**
a coefficient obtained for a storm by dividing the maximum rainfall at any point within the storm area by the mean in the drainage area or basin.

- **rainfall interception**
interception of raindrops by a canopy of vegetation or by a vegetative residue.

- **rainfall penetration**
the depth below the soil or rock surface reached by a given quantity of rainfall which has passed below such surface. The maximum extent of penetration is down to ground water.

- **ranch farming**
area which is not cultivated but left for livestock grazing of natural vegetation.

- **ranching**
the type of farming in which livestock, particularly cattle, are reared on large-scale farms for sale.

- **random**
the occurrence of an event in accordance with the law of chance.

**▪ random selection**
selection of sample units so that each possible sample has a definite probability of being chosen.

**▪ range grasses**
grasses used for pasture or hay, they are often native to the area in which they are used.

**▪ rate and dosage**
these terms are synonymous. Rate is the preferred term. Rate refers to the amount of active ingredient material (such as 2, 4-D acid equivalent) applied to a unit area.

**▪ rate constant**
numerical constant in rate-of-reaction equation. Physically it stands for the rate of the reaction when the concentrations of all the reactants are unity.

**▪ rate meter**
a counting rate meter used to determine continuously the time rate of events such as ionisation .

**▪ rate of flow**
the quantity of water flowing past a designated point within a specified time, such as the number of gallons flowing past a point in 1 minute - abbreviated as GPM.

**▪ rate of flow controller**
an automatic device that controls the rate of flow of a fluid.

**▪ rate of flow recorder**
a recorder indicating the rate of flow of water or other fluid.

**▪ rate of growth**
the rate at which a tree has put on wood, measured radially in the trunk or in timber cut from trunk, measurement based on number of annual growth rings per inch. The rate at which a young animal puts on height and weight.

**▪ riparian**
the doctrine of water right according to which a landowner along a stream is entitled to have the water flow by natural channel undiminished and unpolluted.

**▪ ripening**
acquiring desirable qualities

by being left undisturbed for some time.

- **rock**
  a naturally occurring material composed of one mineral or, more often, a combination of minerals naturally hardened and consolidated into a solid mass. The common concept that material must be hard, as granite or limestone to be classed as rock is not correct from a geological point of view.

- **rodoorn**
  soils containing clay with much iron hydroxide.

- **root**
  descending portion of plant, fixing it in soil, and absorbing water and minerals, and having a characteristic arrangement of vascular tissue.

- **root borer**
  a larval form or insect which bores into the roots of plants.

- **root cap**
  thimble-shaped mass of cells covering and protecting the growing tip of a root. It is formed from the properistem, dermatogen or from the calyptogen.

- **root cell**
  a koutline cell which has cells below it at lower outline levels. All such cells share the same root cell.

- **root climber**
  a plant which climbs by roots developed from stem.

- **root collar**
  the transition zone between stem and root, usually recognised in plants by the slight swelling seen.

- **root crops**
  crops grown for their enlarged roots. The root crops include sugarbeets, mangels, carrots, turnips, rutabagas, sweet potato, cassava, etc.

- **root exudates**
  substances formed in the roots and rejected by them outside, into the soil.

- **root hair**
  thin hairlike outgrowth of an epidermal cell just behind the tip, absorbs nutrients from the soil. They occur in greatest abundance just behind the root tip.

- **root pressure**
  a pressure developed in the roots of some plants which, if the shoot is cut off, causes fluid to exude from the root stump by way of the stele, but the tissues involved and the cause of the pressures are uncertain.

**root reserves**
the storage of plant food in the form of proteins, carbohydrates, etc., in the root.

**root swelling**
the thickened basal part of a tree stem.

**S**
an abundant tasteless odorless multivalent nonmetallic element, best known in yellow crystals, occurs in many sulphide and sulphate minerals and even in native form.

**s curve hydrograph**
a graph showing the summation of the ordinates of a series of unit hydrographs spaced at unit rainfall duration intervals. It represents the hydrograph of unit rate of rainfall excess continued indefinitely. If the ordinates of the S-curves are expressed as a percentage of the total unit hydrograph volume and the abscissa in unit of time or in percentage of lag time, the S-curve is known as a summation graph.

**salminen probe**
a probe for taking soil samples. It is made of a stem one metre long, pointed with a handle, with a cavity 30 cm long in the lower part, one of the edge of the cavity is sharpened and slightly protuberant, it is sunk into the soil and turned, then removed, the soil sample lies inside.

**sample**
a segment of the population selected for marketing research to represent the population as a whole.

**secondary treatment**
waste water treatment, beyond the primary stage, in which bacteria consume the organic parts of the wastes. This biochemical action is accomplished by use of trickling filters or the activated sludge process.

**seed bed**
1. in natural regeneration, the soil or forest floor on which the seed falls.
2. in nursery practice, a prepared area in which seed is sown.

**seed coat**
protective outer layer of seeds of flowering plants. It provides necessary protection to the embryo enclosed in the seed.

**selective absorption**
the ability of a cell, or the cells of an organ, to accumulate a particular ion in a higher concentration in its vacuole, than

it is in the surrounding medium, or alternatively not to accumulate a substance which is present in the surrounding medium, and would normally pass into it by diffusion.

- **self absorption**
the absorption of radiation by the radioactive substance itself. This phenomenon presents a particular problem in the measurement of weak radiation in a bulky sample. This process tends to decrease the observed activity of a sample.

- **self fertilisation**
fertilisation of the egg cells of a plant by the sperm cells of the same plant.

- **self scattering**
the scattering of radiation by the material emitting the radiation. This process tends to increase the observed activity of a sample.

- **semi-perched**
a term applied to ground water or to a water table which has a greater pressure head than an underlying, body of ground water, from which it is not, however, separated by any unsaturated rock. The semi-perched ground water belongs to the same zone of saturation as the underlying body of ground water and where it occurs there is only one water table, which is called a semi-perched water table.

- **semi-perched water table**
the upper surface of a body of free ground water having a greater pressure head than an underlying body of ground water not separated from it by unsaturated material and lying within the same zone of saturation. Semi perched and semi-permeable or impermeable confining beds support perched ground water.

- **semi-permanent pastures**
pasturage of perennial species to be grazed for only a limited number of years, to be re-established after cropping.

- **semi-permanent soil sterilant**
a chemical which makes, the soil sterile for more than 4 months and less than 2 years.

- **semi-permeable membrane**
a physical barrier which only

allows certain particles (based on size and charge) to pass through it, membranes of organelles and cells are semipermeable membranes through which small, uncharged molecules such as water, oxygen and carbon dioxide pass freely.

- **smonitza**
  hydromorphic black or dark grey soil of Yugoslavia usually derived from calcareous clay overlying sand. The surface is leached of $CaCO_3$.

- **sod**
  surface layer of ground containing a matt of grass and grass roots. It holds together and provides a complete and continuous cover of the soil surface.

- **sod bound**
  a term used to describe the unproductiveness of grass sod due to lack of nitrogen.

- **sodium nitrate**
  most of the sodium nitrate used for fertiliser is mined and processed in Chile. It is also manufactured by reacting salt and 70% $HNO_3$ with heating.

- **soil**
  1. a thin layer of earth's crust made up of disintegrated and decomposed rocks, and containing complex mineral compounds, organic matter, water, air, and living organisms. a natural body, synthesised in profile form from a variable mixture of broken and weathered minerals and decaying organic matter, which covers the earth in a thin layer and which supplies, when containing the proper amounts of air and water, mechanical support and, in part, sustenance for plants.
  2. a natural body differentiated into horizons of mineral and organic constituents, usually unconsolidated, of variable depth, which differ from the parent material below in morphology, physical properties and constitution, chemical properties and constitution and biological characteristics. the more or less loose and friable material in which, by means of the roots, plants may or do find a foothold and nourishment, as well as other conditions of growth.
  3. the upper weathering layer of the solid earth crust.

- **soil atmosphere**
  this exists in pore spaces not occupied by water. Gas which occupies the pores of the soil.

**▪ soil colour**

the colour of soil, varying from black, dark brown, red, reddish brown, yellow, bluish green, greenish grey, and grey to white, often indicates important aspects of soil condition. Surface soil generally darker than sub-soil. Generally speaking, dark brown to black colour denotes high content to humus and the associated good structure and adequacy of calcium and nitrogen, soils in depressions and low spots which usually remain wet in humid regions also are black, red or brown colour of the laterites closely associated with good aeration and drainage, in well-drained soil ferric compounds of iron are commonly formed giving it red or brown colour, in imperfectly drained soils iron oxides are formed giving the soil a yellow colour, in humid regions, red or yellow soils can be turned brown by incorporation of large quantities of organic matter, sandy deserts have natural grey colour of their parent rocks, reduction compounds of iron gave a grey or greyish blue colour, accumulation of alkali soils makes soils white or black, depending upon the nature of the salts.

**▪ soil compaction**

is the hardening of soil of more than 5 cm depth. Soil compaction restricts root penetration deep into the soil resulting in reduced uptake of water and plant nutrients.

**▪ soil complex**

an inmate mixture of categorical units grouped into a cartographic unit. It is a soil association composed of such an inmate mixture of areas of soil series and 'types of phases that they cannot be indicated separately on maps the scale used so that the association is mapped as a unit. It is thus a cartographic unit and not a category in taxonomic classification of soils.

**▪ soil compressibility**

the capacity to be compressed under the effect of rains.

**▪ soil degraded**

a soil that has been subjected to stronger leaching than type soils of the area, hence lower in fertility.

- **soil discharge**
direct discharge of ground water through evaporation. The water discharged is, for the most part, lifted by capillarity from the zone of saturation nearly to the surface, where evaporation takes place.

- **soil division**
a grouping of soil series based on both the profile and the mineralogical and chemical composition.

- **soil drench**
to soak or wet the surface of the ground with a pesticide chemical. Generally, fairly large volumes of the pesticide preparation are needed to saturate the soil to any depth.

- **soil fertility**
the quality that enables a soil to provide adequate nutrients in a proper balance for specified plant growth, other factors such as light, moisture, temperature and the physical condition of the soil being favourable.

- **soil formation**
mechanical distintegration and chemical decomposition of parent rocks to form soil.

- **soil formation factors**
the independent variables that define the soil system. Five main groups of soil-formation factors are generally recognised by soil scientists, parent material, climate, organisms (plants and animals), topography, and time.

- **soil fumigant**
a pesticide that will evaporate quickly. When added to the soil the gas formed kills pests in the soil. Usually a tarpaulin, plastic sheet, or layer of water is used to trap the gas in the soil until it does the job.

- **soil genesis**
1. is the result of five factors: climate, biosphere, parent material, topography and time.
2. at any one time and place, the soil is the result of the interacting natural forces, climate and biosphere operating over time on the passive factors of parent materials and their topography.

- **soil geography**
the distribution of main types of soils on the earth's surface. Governed principally by the nature of the rocks by the climate.

- **soil groups**
uniform nomenclature given

to soils possessing certain common physical properties which can be readily recognised in the field like texture, structure and colour of the surface soil.

- **soil injection**
mechanical placement of the fertiliser or pesticide beneath the soil surface with a minimum of mixing or stirring. Common method of applying liquids which change into gases.

- **soil injector**
small hand-operated machine for fumigating soil 5 to 7 inches deep to control soil pests, particularly root-knot eelworms, covers about an acre a day.

- **soil material**
it consists of skeleton grains and plasma.

- **soil maturity**
the ultimate stage of development of a soil from the raw parent material when it is able to express the fluff effect of environmental factors.

- **soil mechanics and engineering**
a sub specialisation of soil science concerned with the effect of forces on the soil and the application of engineering principles to problems involving the soil.

- **soil microbiology**
that branch of science that deals with the study of micro-organisms in the soil.

- **soil mineral**
a natural inorganic compound usually having definite physical properties, crystalline structure and chemical composition.

- **soil mineral matter**
mineral matter contained in soil, corner from rocks from which formed, different soils contain different minerals, according to the chemical composition of parent rock.

- **soil mixture**
1. a mixture of soil and such material as manure, peat moss, leaf mould, sand, etc.
2. for growing seedlings, pot plants and as medium for rooting cuttings. Also called potting soil.

- **soil modification**
the modification of the soil by various operations, apart from the usual cultivation and fertiliser programme, these operations include installation of drains in waterlogged, swampy and meadow soil, removal and distribution of subsurface and subsoil hard-

pans, soil inversion by very deep ploughing or trenching, and the building of terrace.

- **soil moisture capacity**
  the amount of water, expressed as percentage of dry weight, that a soil can retain against the pull of gravity.

- **soil moisture characteristic curve**
  the graphical presentation of the relationship between matric suction and moisture content. In the low suction range (0 to 1 bar), the amount of water retained by the soil depends mainly on the pore size distribution which is a function of the structure of the soil matrix, but in the higher suction range the moisture retention is due mainly to the adsorption phenomena which is a function of soil texture.

- **soil moisture stress**
  the sum of the soil-moisture tension and the osmotic pressure of the soil solution. It is the suction or negative pressure to which water must be subjected to be at equilibrium through a semipermeable membrane with the solution in soil.

- **soil penetrometer**
  is an instrument for recording the force required to push a blunt-pointed probe into the soil. Other conditions being equal, soils with good structure for most agricultural plants are penetrated more easily than soils with poor structure.

- **soil pores**
  interstices (voids) between soil particles.

- **soil salinity**
  the amount of soluble salts in a soil, expressed in terms of percentage, parts per million, or other convenient unit.

- **soil slip**
  the downward movement of soil on a hill side under conditions of complete saturation. smaller in dimensions that landslides.

- **soil slope**
  an incline of a land surface.

- **soil solution**
  the aqueous solutions of salts and other materials, occupying the spaces between the soil-particles.

- **soil stabilisation**
  immobilisation of a unstable soil or of a soil subject to erosion by the natural or artifi-

cial installation of a vegetation.

- **soil statics**
a soil in equilibrium, at least temporarily.

- **soil sterilant**
a chemical that prevents the growth of plants, microorganisms, etc., when present in soil. Soil sterilisation may be temporary or relatively permanent, depending on the nature of the chemical being applied.

- **soil sterilisation**
incorporating a material in the soil which is toxic to plants, the use of steam or direct heat to the soil, particularly in greenhouses, to rid soil of all plant diseases organisms and insects.

- **soil tilth**
strictly the structure of the surface layer of the soil resulting from tilling operations. A soil has a good tilth when it is composed of crumbs which are neither too sticky when wet, not too hard when dry. A good natural tilth may also be brought about in nature, as in certain forest and prairie soils, by organic agencies.

- **soilage**
freshly cut green fodder fed to animals confined in an area.

- **soiled**
pertains to a product which has dirt on or in it.

- **soiling**
the cutting and bringing in of forage for livestock confined to a place in contrast to allowing them to ear in the place where it grows.

- **soiling crops**
crops harvested when in green and succulent condition and fed to animals shortly after cutting. It is neither dried nor stored for future feeding.

- **soloth soils**
an intrazonal group of soils having a thin surface layer of brown, friable soil over a gray leached horizon resting on a brown or dark brown horizon. They develop in a semi-arid or sub-humid climate under shrubs, grasses, or a mixture of grass and trees.

- **step terrace**
terrace with vertical drop on the downhill side, with the flat part suitable for cultivation.

- **step up transformer**
a transformer for increasing an alternating voltage.

**steppe climate**
a climate associated with deficiency of rainfall and high evaporative demand. Intensive agriculture is not possible without irrigation. The high solar radiation associated with steppe climate is favourable to high productivity if irrigation facilities are insured.

**stomatal transpiration**
transpiration of outward diffusion of water vapour into the atmosphere through the pores of stomata and lenticula, the water having been previously evaporated from imbibed membrane into gas spaces within the plant.

**strain**
a group of plants within a variety having very little differences in morphology but having distinct physiological characteristics like uniformity in maturity and having additional desirable quality like higher yield.

**stratified random sample**
a sample from a stratified population consisting of a random selection of sampling units from each stratum, the sampling fraction need not be the same for all strata.

**stratum**
a sedimentary bed or layer that generally consists of the same kind of soils or rock material.

**strongly differentiated soils**
are soils with large differences in the clay contents of their a- and b- horizons.

**strophiole**
one of the small excrescences arising from various parts of a seed testa, never developed before fertilisation.

**structural predisposition**
a pre-existing aptitude to take on a certain structure rather than another. Under normal conditions, every soil acquires a certain structure which depends on its grain-size constitution and on its humus content.

**structural unemployment**
the loss of jobs resulting from structural changes in the economy.

**structure factor**
the ratio of the difference between the total clay obtained by complete dispersion and the clay obtained by water dis-

persion expressed as a percentage of the former.

▪ **structure index**
any measurement of a soil physical property such as aggregation, porosity, permeability to air or water or bulk density, that denotes or indicates the structural condition of a soil.

▪ **structured soil**
soil having a large proportion of aggregates facilitating air and water circulation and the life of microorganisms.

▪ **structureless soil**
that condition in which there is no observable aggregation or no definite orderly arrangement of natural lines of weakness. Massive if coherent, single grained if noncoherent.

▪ **stub**
a short blunt projection, as a blunt projecting stem, branch or root.

▪ **stubble**
material consisting of seed coverings and small pieces of stem or leaves that have been separated from the seeds.

▪ **stubble crop**
crop developed from the stubble of the previous season.

▪ **stubble manure**
farmyard manure, excreta of livestock intermixed with straw and other bedding material.

▪ **stubble mulch**
the stubble of crops or crop residues left essentially in place on the land as a surface cover before and during the preparation of the seed bed and at least partly during the growing of a succeeding crop.

▪ **stubble mulching**
the practice involving direct drilling of seeds after the previous cereal crop has been harvested and stubbles left in the field. It reduces the possibility of soil erosion.

▪ **stump**
the base of a tree and its roots left in the ground after felling.

■ **stump to**
1. to cut back a seedling to make a root and shoot cutting for planting. stunted.
2. an unthrifty plant reduced in size and vigour due to unfavourable environmental conditions, may be due to a wide range of parasitic and nonparasitic agents.

■ **subaerial**
growing just above surface of ground.

■ **subculture**
a culture of bacteria or fungi prepared from a pre-existing culture.

■ **subsoil fertilisation**
the placement of fertiliser at a depth of 40-60 cm behind a subsoiler.

■ **subsoil placement**
placement or fertilisers in the subsoil with the help of heavy power machinery usually followed in humid and sub-humid regions.

■ **subsoiler**
a plough without mouldboard used to work the subsoil, the single, very resistant tool is sometimes shaped like a torpedo, and advances horizontally.

■ **substitution elasticity**
the extent to which one commodity can be substituted for another as a consequence of a given change in their price ratio, if the consumer is to continue to enjoy the same amount of satisfaction.

■ **subsurface irrigation**
consists in laying porous tile pipes in the subsurface below the normal depth of tillage operations. A network of such pipes is made and water is allowed to run in them. The soil sucks this water according to needs. The advantage of this method is that the losses of soil water due to evaporation are prevented.

■ **subsurface tillage**
tillage with a special sweep-like plough or blade which does not turn the surface cover but instead leaves the crop residues on the surface.

■ **subsurface water**
the water of precipitation which penetrates into the soil and occupies the depth between the ground surface and the lower limit of rock-bearing formations.

■ **sulphofication**
the transformation in the soil of organic or elemental sulphur, under the action of microorganisms, into sulphuric

acid, which with the bases of the soil, forms sulphates.

- **sunscald**
localised injury to bark and cambium often resulting in wounds and caused by exposure of a stem to intense sunlight and high temperature.

- **surface tension**
the property, due to molecular forces, by which all liquids through contraction of the surface tend to bring the contained volume into a form having the least area. If an ink is to be compatible with a substrate, the surface tension of the ink must approximate that of the substrate surface. Measured in dynes/cm.

- **taiga**
a broad northern belt of vegetation or biota dominated by gymnosperms, also a similar biota in mountains just below the zone of alpine biota.

- **tail**
the part of the area of a frequency curve lying above or below any given value of the variable shown on the abscissa, as generally used, includes the area at one or both extremes and equals less than half the total area under the curve.

- **tailings**
partly threshed material that has passed through the coarse shakers or straw walkers of a threshing machine, and has passed over the fine sieve.

- **talati**
a village accountant in charge of land records. This term is used in Gujarat, Karnataka and Maharashtra states of India.

- **temperature coefficient**
a factor, which describes the change in a magnetic property with change in temperature. Expressed as percent change per unit of temperature.

- **temperature constant**
in germination testing, a specific temperature which should not vary by more than 1°C.

- **tensometer**
a portable machine for measuring the tensile strength and other mechanical properties of materials.

- **terrace**
a raised bank of earth having vertical and sloping sides. Terraces are constructed to follow the contour of the land and to enable rain and snow water to remain on the land rather than run down slopes and cause soil

erosion. Terraces are a conservation practice to keep soil and water on the land.

- **terrace channel**
the depression located along the upper side of the terrace ridge in which the water flows to the outlet.

- **terrace cultivation**
the growing of crops on level steps or terraces that have been constructed on hillsides. Banks or walls are built to hold the soil in place and, where paddy is grown, to enable water to stand on the fields. Terrace cultivation takes place where land is scarce.

- **terrace interval**
distance measured either vertically or horizontally between corresponding points on two adjacent terraces. The vertical interval (difference in elevation between stake lines) is used almost exclusively in staking terrace lines for construction. For the top or first terrace, it is measured between the top of the hill and the centre of the terrace.

- **terrace system**
a complete series of terraces occupying a slope and discharging run-off into one or more outlet channels.

- **terrace width**
the combined width of the terrace channel and ridge as measured horizontally from the high water line on the upper edge of the channel to the foot of the lower slope of the terrace ridge.

- **terracing**
the construction of level or sloping platforms of earth across a slope to detain or control surface run-off and to check soil erosion.

- **terrestrial soil**
soil developed on land under relatively dry conditions, consisting mainly of mineral of inorganic matter.

- **tertiary treatment**
waste water treatment beyond the secondary, or biologic, stage that includes removal of such nutrients as phosphorus and nitrogen and a high percentage of suspended solids. Tertiary treatment, also known as advance waste treatment, produces a high-quality effluent.

▪ **test of homogeneity**
a test of the hypothesis that there is no real difference between the parameters corresponding to two or more statistics and that differences in the observed values of the statistics are ascribable to chance fluctuations in random sampling.

▪ **test of independence**
any statistical test of the hypothesis that two or more variables are unrelated *e.g.* Chi-square test.

▪ **test of normality**
a test of a set of observations to see whether they could have arisen by random sampling from a normal population.

▪ **texture**
the structure and arrangement of the parts which make up the whole, arrangement of threads in a textile fabric, tissue, structure.

▪ **theoretical velocity**
the velocity which water or other liquid, under a given head, would attain in passing through an orifice, conduit, or other structure were its flow not reduced by friction or other losses.

▪ **theory**
supposition explaining something, especially one based on principles independent of the phenomena to be explained, opposite to hypothesis, speculative view, exposition of the principles of a science, collection of results designed to illustrate principles of a subject.

▪ **thermistor**
an electrical component, usually a semi-conductor, whose electrical resistance changes rapidly with temperature. Used as a sensitive temperature measuring device.

▪ **thermobiology**
study of the effects of thermal energy on all types of living organisms and biological molecules.

▪ **turf**
surface layer of ground containing a matt of grass and grass roots.

▪ **turn under**
ploughing to bring the lower part of the soil to the surface

which covers the former surface.

- **turning**
  the process of moving grain through the air from one bin or storage structure to another or back to the same storage, used in connection with conditioning of harvested produce.

- **turning point**
  a point whose elevation is determined by subtracting the foresight from the height of instrument in levelling. After the elevation of the turning point is known, the instrument can be moved, thus its name.

- **turnout**
  contrivance used for turning out water from a lateral irrigation channel into a ditch, or from a ditch to a field, may be portable or built-in, and is provided with gates. Installed in lined ditches at the time of construction.

- **tyndall effect**
  the phenomenon in which light is scattered by very small particles in its path, it makes a beam of light visible, the scattered light is mainly blue Tyndall effect is useful in distinguishing a colloidal suspension from a true solution.

- **tyrosine**
  an amino acid that is the precursor to the neurotransmitters norepinephrine and dopamine, which regulate mood.

- **U**
  index for an abnormal horizon, *e.g.* a line of pebbles in a B-horizon.

- **U**
  1. symbol for the specific surface area of the soil.
  2. symbol pertaining to subtypes Ui weakly evolved.

- **$U_2$**
  moderately evolved.

- **$U_3$**
  completely evolved.

- **$U_4$**
  1. degraded.
  2. symbol for uniform brown soil.

- **ultimate particles**
  soil particles after a standard dispersing treatment, with all aggregates broken down.

**▪ ultimate ph**
the pH after all diffusible anions and cations have been removed.

**▪ ultimate wilting point**
the moisture content of the soil at which all the leaves of plants growing in that soil are completely wilted and will not recover in an approximately saturated atmosphere without addition of water to the soil. The lower end of the wilting range.

**▪ ultisols**
soils of humid areas characterised by the presence of either an argilic horizon or a fragipan each of which is less than 35 per cent saturated with bases. They have no spodic horizon or no toxic or nitric horizons. They include soils formerly classified as Red-yellow podzolics, Reddish brown laterites, and Rubrozems.

**▪ ultracentrifuge**
a high-speed centrifuge that can attain speeds up to 60,000 rpm centrifugal fields—up to 500,000 times gravity and thus is capable of rapidly sedimenting macromolecules.

**▪ ultramicroscope**
light microscope that uses scattered light to show particles too small to see with ordinary microscopes.

**▪ ultraviolet (uv)**
a band of the electromagnetic Spectrum between the visible and the X-ray, i.e. light that is so blue humans cannot see it. Photons of ultraviolet light are more energetic than photons of visible light.

**▪ umberic epipedon**
a surface horizon like the mollic epipedou but is less than 50% base saturated and not naturally dry more than 3 months per year.

**▪ umberic horizon**
a thick dark-coloured surface horizon (epipedon) similar to a mollic horizon but with low base saturation and may be highly H saturated.

**▪ unavailable**
refers to compounds or chemical elements in the soil or feed which are not in the form available for the plant or animal to use.

**▪ unavailable soil moisture**
soil moisture held so firmly

by molecular forces that it cannot ordinarily be absorbed by plant roots with sufficient rapidity to produce growth.

- **undulate**
to have a wavy form or surface, bend with successive curves in alternate directions, specifically in an up-down direction.

- **unit water requirement**
the weight of water actually used by plants in producing unit weight of dry matter.

- **unpolished rice**
the whole rice grain whose husk has been removed.

- **unregulated market**
a market administered by the local bodies like municipal committees or notified area committees which frame byelaws with the approval of the State Government.

- **unsaturated flow**
the movement of water in a soil which is not filled to capacity with water.

- **unsaturated soil**
a soil that has lost its cation saturation. In which all the fixed cations are $H^+$ ions, the metal cations having left the absorption complex during the exchange. Thus, without exchangeable bases, impoverished, pH 3.0 to 4.0, Usually a soil is partially unsaturated.

- **unstable equilibrium**
in an unstable equilibrium the system returns to a new equilibrium after disturbance.

- **unstable humus**
the humus that has been readily decomposed corresponding approximately to that soluble in 80 per cent sulphuric acid.

- **un-steady flow**
the change brought in with the time in the flow characteristics at the channel cross section. Intermittent runoff into terrace outlets, diversion channels, etc., is generally unsteady.

- **upland**
a high raised land.

- **upland rice**
the type of rice that is grown without irrigation.

- **upland soil**
the area of higher altitude that is elevated about the lowland along rivers or between hills.

- **upper critical limit**
the concentration of a nutrient element in plant or soil above which the sufficiency range sets in.

- **urban forestry**
when groves avenues and all vacant areas within the habitations are created to reduce pollution of factories and dust.

- **urban land**
is land so altered or obscured by urban works and structures that identification of soils is not feasible. Soil boundaries should be extended into urban areas wherever it is possible to do so with reasonable accuracy, and the use of this miscellaneous land type is restricted to the closely built-up parts of cities.

- **urban runoff**
storm water from city streets and gutters that usually contains a great deal of litter and organic and bacterial wastes.

- **urea**
a crystalline colourless nitrogenous compound that can be synthesised from carbon dioxide and ammonia for use in fertilisers, resins, and plastics.
$CO_2 + 2NH_8 \longrightarrow CO(NH)_2 + H_2O$
Urea is a white crystalline solid, very soluble in water. It contains the highest per centage of N of any solid form 45-46%. Since the nitrogen in urea is rapidly converted to ammonia in the soil, urea is considered an ammonical fertiliser. The high analysis means less bags to handle and low cost in transportation.

- **urea sulphuric acid**
a new liquid fertiliser material being manufactured as a reaction product of urea,

sulphuric acid and water $(NH_4)_2CO+H_2SO_4=(NH_2)_2CO$-f $H_2SO_4$+Heat. Depending on the ratio of the reactants, nitrogen content can vary from 28% to 10% N, and sulphur content from 9% to 18%S. A remarkable property of urea sulphuric acid fertilisers is that they virtually eliminate the corrosive action of sulphuric acid on human tissue, with only minimal reduction of the acidifying effect, making it a suitable material for reclaiming sodic soils and for treating irrigation waters as compared to sulphuric acid. The corrosive and hazardous properties of sulphuric acid have often restricted its use. Urea sulphuric acid may cause a mild skin irritation after prolonged skin contact, and the product will sting if it comes in contact with a cut or sore on the skin. However the product will dissolve clothing and deteriorate shoe leather, so it is recommended practice to use rubber protection equipment, especially footgear.

- **urea super granules**
urea granules of 1 to 3 g size developed for deep placement in reduced zone of rice soils, helps to minimise gaseous losses besides ensuring sustained availability of nitrogen over a longer period of time.

- **urea z**
the reaction of urea with acetaldehyde yields a mixture of compounds consisting mainly of ethylene diurea and 2-ethylene-3-urea. Small amounts of ethylolurea and unreacted urea may also *be* present. Urea-Z contains 33-38 %N, of which about 15% is water soluble.

- **ustic moisture regime**
in most years, these soils are dry for more than 90 cumulative days but less than 180 days. In temperature (non-iso) climates, they are usually moist at least 45 consecutive days in the 4 months after the winter solstice and not dry 45 consecutive days in the 4 months after the summer solstice.

- **util**
the equivalent of one unit of satisfaction. ('utility') derived from the consumption of a good.

- **utilisation (forestry)**
is the branch of forestry which deals with the harvest-

ing, marketing, conversion and putting the forest produce for a variety of uses *e.g.* timber, fuel, charcoal, pulp wood, plywood and a host of other products.

- **utilisation coefficient of fertilisers**
  the ratio of the quantity of substance withdrawn by the plant to that applied to the soil in the form of fertiliser.

- **utilisation coefficient of soil's reserve**
  the ratio of the quantity of a nutrient substance present in the soil to that withdrawn by the plant.

- **utilisation value**
  in forestry, expression denoting the usefulness of a seed determined by its purity and germinative energy.

- **utility**
  the quality of being of practical use. Opposite of disutility.

- **utility function**
  the mathematical relationship between the quantity of a good consumed during a given time period and the amount of satisfaction derived, other things being equal.

- **v ditcher**
  a wooden tool used for construction or cleaning of field channels for irrigation or drainage, also used for making border ridges or bunds.

- **valley**
  a long, narrow land area lying between two areas of higher elevation, often containing a stream.

- **van bemmelen factor**
  a multiplication factor for converting the organic carbon content of soil into organic matter. Its value is 1.724 and is based on the assumption that soil organic matter is 58% C.

- **van de graaf generator**
  an electrostatic generator. An insulated electrode is charged to a high voltage by a woving conveyor belt.

- **variable costs**
  are those costs associated with production that changes directly with the amount of production, e.g., the direct

material or labor required to complete the build or manufacturing of a product.

- **variable error**
  the deviation of obtained scores or measures from their respective true values, as a result of the operation of random factors that affect in varying degree and direction individual scores or measures.

- **variance ($a^2$)**
  a measure of dispersion of a set of data points around their mean value. The mathematical expectation of the squared deviations from the mean. The square root of the variance is the standard deviation.

- **variance ratio**
  the ratio of two estimated variances. Twice the natural antilogarithm of z. Sometimes denoted by F.

- **variant**
  a variable quantity that is random.
  a specific numerical value of a variable.

- **variate**
  a variate is quantity which may take any of the values of a specified set with a specified relative frequency of probability.

- **variation**
  1. when there is a difference between individuals due to differences in the permanent structure of their cells, or to differences in the environment.
  2. the differences between the offspring of a single mating.

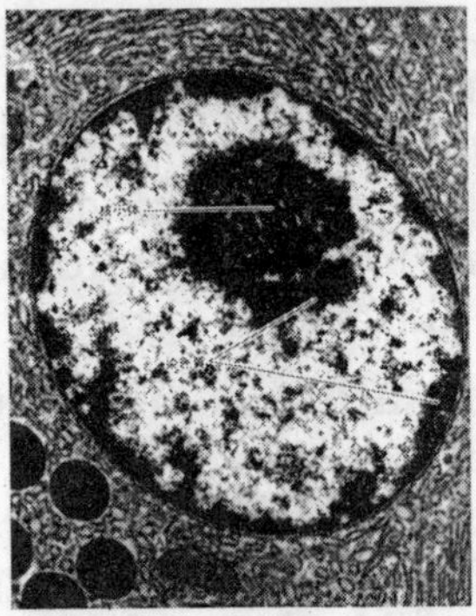

- **varieties (soil)**
  this category differentiates according to texture at a lower level of generalisation than used at the level of the genera.

- **variety**
  1. population of plants having many characteristics in common, a variety may be a pureline, a mixture of pure lines, a Mendelian population of a clone.
  2. a category of things distinguished by some common characteristic or quality.

- **varnish, desert**
  a glossy coating of dark-coloured compounds, probably composed mainly of iron oxides, covering pebbles,

stones, and large rock surfaces exposed in hot deserts.

- **vegetable forcing**
  the growing of vegetables ahead of their normal growing season, producing vegetables at a time when their supply is limited.

- **vegetable oil**
  oil obtained from the seeds or nuts of vegetable growth. Some of these are "drying oils," such as linseed, soya, tung and oiticica, which are used as binders for oil-based paints and varnishes.

- **wage goods**
  a term first used by Prof. A. C. Pigou for the goods, the prices of which determine the value of real wages as distinct from nominal wages.

- **wage minimum**
  a basic payment, established by law, by agreement or by action of an employer which is not to be lowered for workers in a given area.

- **wage rate**
  determination of wages on a specified time or unit basis, *i.e.*, per hour, per piece etc.

- **water compaction**
  water furnished by destruction of pore space owing to compaction of sediments.

- **water free space**
  the sacs in a tissue in which salts move by diffusion so that their concentration equals that of the external medium.

- **water furrow**
  a shallow trench or depression between two raised soil beds in which irrigation water or surface drainage water can flow.

- **water intake**
  the movement of irrigation water from the soil surface into and through the soil. It is the expression of several factors, including infiltration and percolation.

- **water level**
  the elevation of a particular point or small patch on the surface of a body of water above a specific point or surface, averaged over a period of time sufficiently long to remove the effects of short period disturbances.

- **water of compaction**
  water furnished by destruction of pore space owing to compaction of sediments.

- **water table gradient**
  rate of change of altitude per unit distance in the water table at a given place and in a given direction.

- **water table perched**
  the upper surface of a body of free ground water in a zone of saturation separated, by unsaturated material, from an underlying body of ground water.

- **water table profile**
  1. the line formed by the intersection of the water table with a vertical plane.
  2. the term is also applied to a graphical representation of such a line.

- **watershed**
  the specific land area that drains water into a river system or other body of water.

- **watershed area**
  a topographic area within a line drawn connecting the highest points uphill of a drinking water- intake into which overland flow drains.

- **watershed management**
  the management of all the natural resources of a water-

shed to protect, maintain, or improve its desired water budget, both quantity and quality, over time.

- **watertable**
the top of the groundwater. A swamp or a lake in a low-lying area might be caused by the watertable being higher than the low ground
- **weather map**
a map showing the principal meteorological elements at a given time and over an extended region. Meteorological elements noted include the barometric pressure, the temperature, the state of the weather, areas of precipitation, and wind conditions.
- **weed eradication**
the elimination of all live plants, plant parts, and seeds of a weed infestation from an area.
- **weed tree**
1. a tree which grows where it is not wanted.
2. an undesirable species of tree which interferes with the development of crop trees.
- **weeder**
a hand tool for removing weeds such as spring-tooth harrow, hand-hoe, rotary hoe, *khurpi*, etc.

- **weedy**
an abundance of weeds in a field, a growth of high vigour in a plant, usually associated with fruiting.
- **wet soil**
1. a soil which is wet or inadequately drained, for short or long periods of time.
2. marshy or boggy soil.
- **wet spell**
a duration of at least 15 successive wet days.
- **wicn effect**
an increase in the conductance of an electrolyte at very high potential gradients.
- **working farm capital**
the livestock, machinery, equipment, feed, other supplies and cash, used in operating a farm business.
- **working plan**
a written scheme of management aiming at a continuity of policy and action and con-

trolling the treatment of a forest.

- **working population**
the proportion of the total population that is either in work or willing to work.

- **workshop**
a place or shed on a farm where machinery or equipment are repaired.

- **wound hormone**
a substance produced in wounded tissue, which is able to influence the subsequent development of parts of the plant.

- **wound parasite**
a parasite which can only gain entry into the body of a plant through a wound.

- **wound tissue**
a pad of parenchymatous cells formed by a cambium after wounding. It may give rise to groups of meristematic cells from roots and buds form.

- **x**
a symbol for frangipane character used as a suffix with horizon designations to indicate genetically developed properties of firmness, brittleness, high density, and characteristic distribution of clay that are diagnostic of frangipanes. Frangipane, or parts of frangipane, may qualify as A2, B, or C. Such horizons are classified as A2, B, or C, and the symbol 'x' is used as a suffix to indicate frangipane character. Unlike comparable use of supplementary symbols, the symbol 'x' is applied to B without the connotative Arabic numeral normally applied to B. Arabic numerals used with C to indicate only vertical subdivision of the horizon precede the 'x' in the symbol, as Cix, C2x.

- **x bodies**
protein-like inclusions in cells affected by a virus.

- **xcrothermique**
the number of days without rainfall, corrected taking into account the days of fog, of dew, and relative humidity according to a certain scale.

- **xerothermic index**
an index of the relationship of drought to plant growth.

- **x-ray crystallography**
a technique for determining the 3-D structure of a molecule, based on the diffraction of x-rays by the crystallised form of the molecule. Computational analysis of the diffraction pattern produces an

electron density map that is optimised using sequence.

- **xylophyte**
a woody plant.

- **yarn irregularity (cotton)**
variation in uniformity. It is an important characteristic of a sliver, roving or yarn, since an even sliver may be expected to give an even yarn which will be stronger and can be more efficiently spun and woven, with fewer breaks or warp streaks.

- **yarn number (cotton)**
a measure of the linear density or fineness of yarn, and denotes the number of 840 yard-lengths in a pound in the case of cotton.

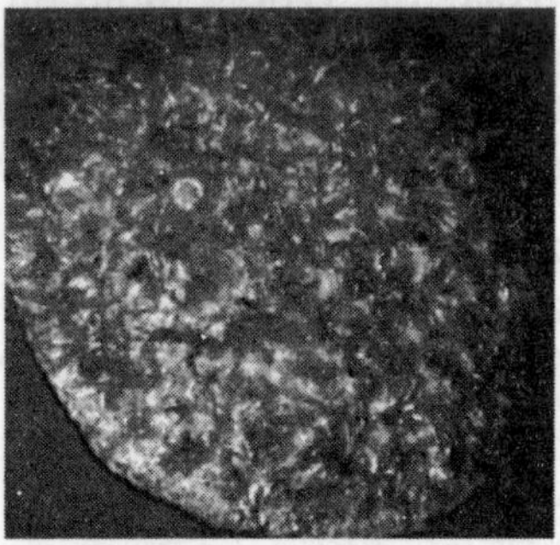

- **yarn twist (cotton)**
the amount of twist in a yarn is generally expressed as the number of turns per inch, but it may also be expressed by the helix angle in a structure of known diameter. The number of turns per inch is determined by counting the turns in a measured length of yarn, as it unwinds in a twist tester.

- **yield value**
stress in an adhesive joint at which a marked increase in deformation occurs without an increase in load: $V = k\,u\,(F-f)$.
where V is the volume of flow, u is the coefficient of mobility, F the force applied, f the force necessary to overcome the cohesive forces of the system and is known as yield value.

- **z**
the natural logarithm of the ratio of the two estimated standard deviations.

- **zaid**
the season between Rabi and Kharif *i.e.* March to June.

- **zamindaar**
a landowner responsible for the collection of taxes during the British rule.

- **zeatin**
a growth-promoting substance, first isolated from maize *(Zea mays)* kernels. A derivative of adenine.

$CH_3$-C—$CH_2$OH.

- **zone of capillarity**
  an area that overlies the zone of saturation and contains capillary voids, some or all of which are filled with water that is held above the zone of saturation by molecular attraction acting against gravity.

- **zone of intermittent saturation**
  a layer of soil that lies just below the surface layer of soil and which will contain ground water after rainfall, but which will tend to dry out during periods of drought. Thus, the zone contains water only intermittently and the frequency of rainfall will determine the periods during which the zone is saturated.

- **zone of saturation**
  the space below the water table in which all the interstices (pore spaces) are filled with water. Water in the zone of saturation is called groundwater.

- **zone of weathering**
  that depth or thickness of the surface crust of earth in which weathering has taken place. It includes soil and parent material of soil.

- **zoogene**
  a soil or horizon one of the primary factors in the formation of which is the action of animals *e.g.*, earthworms, arthropod, etc.

- **zoophyte**
  any of various invertebrate animals resembling a plant such as a sea anemone or coral or sponge.

- **zwitterion**
  the ionic state of an amino acid when the whole molecule is electronically neutral but contains one positive and negative charge. COO.

- **zygo**
  a prefix from the Greek 'zygon' meaning yoke.

- **zz**
  symbol for the coefficient of decomposition of the litter.